OXFORD

WATCH YOUR CHILD'S

Watch your Child's Weight

JENNIFER J. ASHCROFT
*Principal Lecturer in Clinical Psychology,
Lancashire Polytechnic*

and

R. GLYNN OWENS
*Lecturer in Clinical Psychology,
University of Liverpool*

Oxford New York Tokyo
OXFORD UNIVERSITY PRESS
1987

Oxford University Press, Walton Street, Oxford OX2 6DP
Oxford New York Toronto
Delhi Bombay Calcutta Madras Karachi
Petaling Jaya Singapore Hong Kong Tokyo
Nairobi Dar es Salaam Cape Town
Melbourne Auckland
and associated companies in
Beirut Berlin Ibadan Nicosia

Oxford is a trade mark of Oxford University Press

© Jennifer J. Ashcroft and R. Glynn Owens 1987

All rights reserved. No part of this publication may be reproduced,
stored in a retrieval system, or transmitted, in any form or by any means,
electronic, mechanical, photocopying, recording, or otherwise, without
the prior permission of Oxford University Press

This book is sold subject to the condition that it shall not, by way
of trade or otherwise, be lent, re-sold, hired out or otherwise circulated
without the publisher's prior consent in any form of binding or cover
other than that in which it is published and without a similar condition
including this condition being imposed on the subsequent purchaser

British Library Cataloguing in Publication Data
Ashcroft, Jennifer J.
Watch your Child's Weight
1. Obesity in children—Prevention
2. Children—Nutrition
I. Title II. Owens, R. Glynn
613.2'5'088054   RJ399.C6
ISBN 0-19-261645-5

Library of Congress Cataloging in Publication Data
Ashcroft, Jennifer J. (Jennifer Joan), 1953-
Watch your child's weight.
Bibliography: p.
Includes index.
1. Children—Nutrition. 2. Obesity in children—Prevention.
3. Exercise for children. I. Owens, R. Glynn.
II. Title. [DNLM: 1. Children Nutrition—Popular works.
2. Diet, Reducing. 3. Obesity—in infancy & childhood—
Popular works. WD 212 A823w]
RJ206.A84 1987   649'.3   87-11127
ISBN 0-19-261645-5 (pbk.)

Set by Hope Services, Abingdon, Oxon
Printed in Great Britain by
Richard Clay Ltd
Bungay, Suffolk

To Barrie, Chloë, Henry, and Lizzie

Preface

In modern society weight control is a great problem for many people. It is especially sad if children have weight difficulties. They often do not know how to exercise self-control, or they cannot change the type of food that is given to them by others. Parents will be wary, quite rightly, of putting their youngsters on a standard diet meant for adults. Yet help is needed. Children who are an inappropriate weight suffer physically. They may suffer emotionally as well. This book sets out to solve children's weight control problems. It outlines methods which are specifically tailored for youngsters' needs. These methods are completely safe, easy to understand, simple and fun to follow.

Recent reports show that many children eat an unhealthy diet. They have too much sugar, too much fat, and not enough fibre. It is especially important that children with weight problems get this balance right. We show you how to achieve good nutritional goals in a way designed to maximize health. If children eat the right things and exercise in an appropriate way, good weight control should be easily established. Whether the problem is overweight or under-weight, over a period of time small changes to health habits add up to give a good physical shape. We recommend that any changes you make to children's diet be introduced very gradually, in tiny steps that are easy and fun to achieve. That way there is no back-sliding.

Although it might be appropriate to set up a specific programme for a youngster's food intake and for exercise, we emphasize throughout the book that children should be viewed in the context of family life, school life, and social life generally. A successful programme of weight control will not isolate a child from events which entail eating with others. Children tend to like going to parties, having meals with the family, receiving chocolate Easter eggs at Easter, and so on. We show you how to cope with such occasions so that the children stay happy but do not lose control of their weight in the process.

If you know a child with weight problems and you want to help, then this book is for you. It is an essential guide for parents of children with weight problems, and for health professionals concerned with weight control for children.

January 1987
J.J.A.
R.G.O.

Contents

PART 1

What you need to know before starting the weight control programme

1 How to tell if your child has a weight problem 3
2 A quick look at family and friends 14
3 The parents' attitude to eating habits and exercise 27
4 What is a healthy diet? 36
5 The importance of exercise 59

PART 2

The weight control programme

6 It is never too early to start 73
7 How to change to a healthy eating plan for the whole family 82
8 Setting up a weight control programme with your child: diet 102
9 Setting up a weight control programme with your child: exercise 116
10 What happens once your child's weight is under control? 137

Further reading 141
Index 143

PART 1

What you need to know before starting the weight control programme

1 How to tell if your child has a weight problem

It is the school sports day. All the children are there, some obviously doing well for their team, firm favourites with their friends. In the individual events, too, each child is trying hard, is having fun. There is laughter, yelling, everyone having a good time. Well, almost everyone. How about the chubby child, the one with the plump knees sticking out under the extra large pair of shorts? This child is not doing so well today, huffing and puffing at the back of the field of runners. That child may be popular enough as a friend when not on the games field, but is never picked for the team events.

And, of course, there are some overweight children, who never seem to make it to sports day. They know they look different, are at a disadvantage at sports, and become mysteriously ill on games days. They may develop a headache, sometimes a stomach ache . . . any reason that might be good enough to be excused sport.

Does it matter? Is a little plumpness anything to worry about? The most recent research strongly suggests that it is. The wrong eating habits in childhood can cause many health problems; your child might well become too fat or go to the other extreme and be too thin. If your child is clearly a very different shape from other children, this could lead to a great deal of unhappiness. Friends might taunt or tease.

The fat child is going to be less active, less good at games, not picked for team sports, and so on. In certain situations then, the obese child is going to feel less worthy than others. If such children see themselves as losers at this early stage, what chance do they have when adult?

Apart from this loss of self-esteem, there is also physical damage to consider. So you think a little puppy fat never hurt anyone? How long has your child had this puppy fat? One year? Two years? Five years or maybe longer? It is true that babies and very young children have a slightly rounded look about them. Nobody is suggesting that babies or young toddlers should have hollow cheeks or be stick thin. On the other hand, they should not be a mass of blubber either. The modern trend is to emphasize good nutrition; if possible, breast milk rather than bottled milk for babies, not adding sugar to food or drinks, and so on. The days when a good-looking baby or toddler was as fat as possible are gone. Rather than trying to get a child to be 'big and bonny' we should be encouraging the development of healthy eating habits from as early an age as we can.

Most children who are overweight, most that are underweight, and even many of the children that are average weight, are eating a dangerously unhealthy diet. This might show itself in childhood in abnormal body size, rotting teeth, or teeth full of fillings. Or the developing ill health might not show itself until later in life, when the bad eating habits have built up over the years to contribute to adult obesity, or arthritis, diabetes, or gall-bladder disease, or even cancer, stroke, or heart problems. Many of the illnesses which people associate with old age are strongly related to a faulty diet, bad eating habits first established in childhood and continued through the years.

Parents naturally want the best for their child both now and later, when they have grown up and left home. This book is about giving your child the best start to a happy,

How to tell if your child has a weight problem

healthy life. So, what is the first step for you to take? As adults we tend to be rather obsessed with how much we weigh. A large number of us are too fat. We go on a diet, lose a little weight, stop the diet, then gain a little weight, and so on. This system does not actually work well for controlling weight in adults. Many of us spend our lives trying diets, never managing to stay slim, healthy, and happy with ourselves. So it seems pretty pointless to start this sort of thing with our children. In fact, it could be downright dangerous to have youngsters follow many of the diets that are available for adults. This, therefore, is the first perhaps surprising thing for you to learn:

- Weight control for children does *not* involve a strict diet, especially not one with drastically reduced food.

So, if you think your child is slightly overweight (or even positively plump!), or even if your child is at the other extreme, as thin as a reed, it is *not* just a case of starting on some standard diet picked from any book you happen to find on the bookshelves—a calorie-restricted diet for a fat child, a diet extra high in calories for a thin child. If you try to do this, you might well damage your child's health. Growing children need good, nutritious food. If you pick, for example, one of the many slimming diets for adults, you will often find milk restricted to half a pint a day, or even less. While this might be fine for some adults, cutting down on milk in a child's diet is a rash and possibly dangerous thing to do. Children need plenty of calcium and milk is a major source of this mineral in the diet. Restricting a child's milk intake could lead to serious illness. Reasons such as this mean that slimming diets for adults are not suitable for children.

But what if you think your child is too thin? Just piling extra food (in the form of fatty or sugary things) on to your child's plate is not going to help. Good nutrition is essential for good health and good physical shape, and this means giving children foods rich in essential nutrients, vitamins, minerals, and so on. You will not find many of these in sticky confections, pastries, chocolates, and sweets.

For maximum health, all children, whether overweight, underweight, or just right, would do well to follow an eating plan which is generally low in fat, low in sugar, low in salt, and high in fibre. If such a plan is going to be followed, it is essential that it be introduced very gradually, in small, easy-to-manage steps. It should be fun to follow. It should *not* mean deprivation and misery. It should fit in with a normal style of life—if your child is given chocolates as a present, or is invited to a party where there are cakes and other sticky things, of course there should be the freedom to eat these things. Eating in a healthy way does *not* mean that chocolate or chips are forbidden for ever. Moderation is the secret. If eating habits are generally good, then the odd bar of chocolate or similar thing is not going to make a great deal of difference. We need to stand back and consider what the overall food intake is like in order to judge whether a diet is health promoting or damaging. Later on we will show you exactly how to do this, but, basically, if the diet is healthy, if the child is doing appropriate exercise, and if these things are made fun, good health and a good physical shape should develop naturally.

Is your child too fat, too thin, or just right?

The extremes of fatness or thinness are usually easy to recognize. Look at your child and make a comparison with others of the same age and height. Is there an obvious difference? Perhaps overweight is more easily recognized than underweight. Most children tend to be relatively slim, and it might be difficult to tell if your child is thinner than the average. You can always do what most of us have done at some time and resort to charts of suggested weight. Most

How to tell if your child has a weight problem

adults have, at some point, looked at tables which give the average weight of adult males or females. This only gives a rough guide to what you should weigh. If you are small-boned you are supposed to weigh less than someone of the same height who is big-boned, and so on.

Should we ask our children to do the same as we do? Should we get them to pop on the scales while we find a chart and work out the problem in terms of an exact number of pounds (or kilograms) over or under the average for a child of the same height? We can have a go at this, and in fact we give you a chart for children later in the book, *but* this is only a *very* approximate guide to the desired weight. Your particular child might be slightly over the suggested weight but be very muscular rather than fat, very healthy, and in good shape. Alternatively, your child might be a slightly lower weight than average but be very small-framed and perfectly fine. Do not be too fussy about your child's actual weight. Be more concerned with physical shape.

Setting goals for weight change for children is a very different thing to setting them for adults. Suppose, for example, that an adult woman of 5′ 4″ (1.625 m) is 7 lb (3.2 kg) overweight, according to height/weight charts. Within one year she loses this extra weight and therefore becomes about right for someone of 5′ 4″ (1.625 m). Now let us take a child of 4 ft 6 in (1.37 m) who is also about 7 lb (3.2 kg) over the average weight for someone of this height. If, one year later, this child has lost this extra 7 lb (3.2 kg) then the correct weight will have been reached for a person who is 4 ft 6 in (1.37 m). *But*, of course, children grow, and it could be that this child has shot up over the past year and is therefore now *underweight*. So it is inappropriate to set goals for weight for children without considering that their height will increase. In fact, if an overweight child stays the same weight, because height is increasing, eventually physical appearance will change. This year's shorter, fatter shape can be next year's tall, slender figure! Therefore, we need to be very careful about setting goals for a child's weight control. We go into some detail over this problem later on.

The importance of tackling the problem straight away

If there is something wrong with a child's weight, the chances are that there is also something wrong with the diet. Actually, most children, regardless of weight, could benefit from an improvement in diet. There is no point in waiting for a year or more, hoping a weight problem will sort itself out if the general diet is wrong. It is important that children get good nutrition for healthy development. This does not mean cranky diets or very expensive foods, but sensible eating. It is not fair to your child to wait, fingers crossed, in the hope that things will turn out all right. By taking direct action, that is by weeding out bad eating habits, you will automatically be going a long way towards good weight control as well.

If you decide not to take any action, the weight problem *might* just sort itself out. We have all seen children who seem plump one year and then fine the next. On the other hand, the problem might not just disappear. It might get worse. The fat child might get fatter, the thin child thinner. And even in cases where a weight problem does seem to go of its own accord, if the basic diet is still bad there will be problems again at some time. A diet too high in fat and too high in sugar is likely to lead to obesity or very serious illness at some time in the future.

So, the time to act is now. You must make up your mind

How to tell if your child has a weight problem

that you want to help your child. If you are going to do this effectively, you need to know about three basic things. First, you need to find out about the best kinds of foods to eat for maximum health. Secondly, you should know a little about the most appropriate exercise for children. Thirdly, and most importantly, you will need to know how to make these changes easy and fun for you and your child. These are the things we will show you in this book. We would like to bet that your youngster has fads and fancies for certain foods, and that these foods are probably either high in fat, high in sugar, or high in fat *and* sugar. We all know plenty of children who love chips, chocolate, cake, or crisps. Not so many are really crazy about fish, sprouts, or cottage cheese! Therefore, it is really important that any changes to the diet are made very slowly and carefully. Take the path of least resistance. Sudden, large alterations are likely to provoke an equally sudden and violent reaction from your angry offspring! Foods that are especially liked should not be totally banned. You will be developing new tastes in your child and this is bound to take time. You can expect it to take quite a few months, even a year or two, before all the necessary changes to a good way of eating have been made. By then you will have built up healthy habits and healthy tastes which are much more likely to be continued. Taking a year to change eating habits in childhood is really very little time when you think that the benefits are likely to continue throughout life. Establishment of a good lifestyle cannot, and should not, be hurried.

If your child has a weight problem, you should see this gradually go as you progress with the programme of diet change. Exercise, of course, is very important too. Children are usually very active, but those who are overweight might just fight shy of sports and the like. As they become trimmer and healthier, games and exercise should become much more enjoyable.

Make things fun and make things easy

As adults we tend to be remarkably hard on ourselves when it comes to changing habits. New Year's Eve comes along and we make ourselves a list of resolutions: I will give up smoking; I will give up eating cakes and chocolates; I will go jogging every day, and so on. Just look out for the number of people you see jogging on New Year's Day. Decked out in new track suits they make a brave plod around the block. Some individuals, of course, stick with these good intentions. They try to change all their bad habits in one go and they succeed. They deserve a medal! By far the majority of people, however, do not succeed. Usually by about mid-January the resolve has gone and the bad habits have crept back in.

Usually, we fail to learn from this experience. Next New Year, here we are again with our list. It is along much the same lines as last year: I will go on a diet; I will go to exercise classes; this is the year I will get slim for ever, and so it goes. Again, come the end of January the diet is broken, maybe one or two of the exercise classes have been skipped, and it is highly unlikely that we will get slim just yet!

There are some very basic lessons to be learned from this, both for yourself and in helping your child with a weight problem. If you set too difficult a task, success just will not happen. If you set too many hard tasks, too many things to change all in one go, this is likely to result in failure. Therefore, it is better to make goals easy to attain by breaking them into small, easy-to-manage tasks or steps. It helps if you do not tackle too many things at once. Do not overload!

Remember, too, that if you are trying to change to a more healthy lifestyle, it helps if you can build in an element of enjoyment. This is especially important for children, where encouragement and rewards for appropriate behaviour can reap dividends. Set them small, obtainable goals and give them a small present, or at least plenty of praise and encouragement, each time a goal is reached. You may like to use bright colourful charts, or award stars to show how progress has been made. You will be surprised at the difference things like this can make. We will be telling you more details about this in later chapters. With care, you can do a great deal to help motivate your child.

Parent participation

It will be difficult for your child to reach the right weight without help from you. After all, it is parents who go shopping for food and decide what will be available in the cupboards and fridge. It is you who prepares the meals. And it is probably you who decides if snacks are readily available, if cakes, sweets or biscuits are always to hand. If a child is thirsty, it is often a parent who offers lemonade or squash rather than water.

As parents you build up certain expectations in your children. If you buy lots of sweet foods and drinks, it is probably not surprising if your youngster expects to consume these things. It is perfectly natural for you to want to treat your child occasionally to the things that children in our society have learned to regard as special. If every other child has a chocolate egg for Easter, why shouldn't yours? This is fine *in moderation*. But there is a limit to what a child should expect to have. Too many of the wrong things, for too much of the time, will surely lead to health problems. This is true for the child with a weight difficulty, but also for the rest of the family as well. It is not so much that overweight is inherited. It is much more the case that inappropriate eating and exercise habits are passed on from generation to generation, and in some individuals this results in obesity, or maybe the more severe diseases we mentioned earlier.

Before using the weight control programme

The good news is that bad habits that have been acquired through a process of learning can be unlearned too. New habits can gradually take their place. But retraining appetite is going to depend upon more than forbidding your child certain foods. General habits relating to eating behaviour might need to be altered as well. For example, say the problem is that your child is slightly overweight, is eating too many of the wrong kinds of food. Is it fair to leave chocolates and sweets scattered around the house and expect your child just to ignore them? Pack these foods out of sight if you must have them at all. There again, if you are training your child not to eat too many sweet things, it is hardly fair if you sit there and, in full view, plough your way through a large box of chocolates! Set a good example as much as you possibly can. It is surprising the extent to which children model themselves on their parents. In the same way, your own behaviour with respect to exercise will be an important influence on your child. If you jump into the car to go to the shops, when you could walk in a matter of minutes, it is not surprising if your child comes to believe that exercise is something to be avoided.

Finally, it is always wise to remember the positive aspects of habit change. You will not just be getting rid of your child's undesirable eating behaviours, you will also be developing a health-promoting lifestyle. You should be experimenting with new and exciting tastes, new recipes for the whole family to try. Also, if you are encouraging your child to exercise, you might find you and other members of the family are also exercising more. For example, you might find yourself walking a lot with your child rather than taking a bus or going by car. You might be more adventurous and all go swimming. The possibilities are endless and it is essential to find something which you can really enjoy. In doing all these kinds of things, you should find that every member of the family will benefit, not just the child with the weight problem.

Summary

It is up to you and your child to decide whether or not weight is a problem. Although it is possible to use height and weight tables for children in the same way as we do to assess adults' weight problems, in general simply looking at your child will tell you whether or not weight control is problematic. If your child does need help, we cannot stress too highly that this *does not* involve putting your child on a strict diet. Rather, you are looking to achieve a generally healthy lifestyle for your child. Any bad habits that your child has learned—eating lots of sweets for example—will need to be unlearned. You will need to consider your child's diet, the exercise your child takes, and how to bring about changes in these in a way that is enjoyable and fun. This will probably mean changes in your own lifestyle as well—but the changes we are talking about are ones which will be positive for you, your child, and for the whole family.

2 A quick look at family and friends

Eating—it is a social thing

When adults try to control their weight they often make drastic, sudden changes to their usual diet. This often makes them a bit miserable. For example, a beer drinker might decide to cut out all alcohol in order to lose excess fat. Someone who is very keen on sweet foods might decide suddenly to omit totally every little bit of sweet stuff. Or, alternatively, many adults pick a totally different way of eating, relying heavily on so-called slimmer's foods or powdered mixes which make up into special diet drinks. But as soon as slimmers start their chosen plan of action they often find themselves beset with difficulties they perhaps did not anticipate. For, as well as hunger, there are also social consequences of a diet. The beer drinker might feel a bit left out drinking slimline tonics if all around are drinking beer as usual. The person keen on coffee mornings or visiting friends for afternoon tea might feel awful about refusing a specially made cake or biscuits. People very often cut down their socializing in order to avoid such situations. And the person on a very rigid diet of special slimming drinks or mixes is going to have even more problems. Family meals could be a nightmare in the end. If all around you are tucking into a big roast for Sunday lunch, and you are sitting there, starving hungry, with a slimmer's milk drink, or a couple of crispbreads and a tomato, how happy do you think you will be? How long will your diet last?

Now, if you go on a diet, doing so is your own decision. If it makes you miserable, it is something you have brought on yourself. Imagine how much worse a child is likely to feel if forcibly placed on a strict diet. There will be hunger. There could well be a certain amount of social isolation (not eating family meals, not being able to eat the food at other children's parties, or maybe missing parties altogether). And, on top of these, there are likely to be feelings of bitterness and resentment towards the parents who have caused this suffering.

The message is clear. Very strict diets or eating regimes which do not fit in with your family and social life do not work. They fail for most adults who use them. It is cruel to subject your child to them. Whatever changes are made, they must not make your child isolated from family and friends.

However, this does not mean that your hands are tied, that you are powerless to help. Rather, it means that you have to be subtle in changing your child's diet to a healthy one; we will show you how. In the long term, a child whose weight is being brought under control is being helped to feel more like the other children, more sociable. If a young person is too fat or too thin, this automatically makes for a feeling of being different. If, over a year or so on a healthy diet, your child's weight and shape become more like those of the other children, you will have done a lot to boost your youngster's self-esteem and, ultimately, general popularity with friends.

Is there anyone else in the family with weight problems?

Take a look at your whole family. Is your child really the only one with weight problems? Research has shown that a child whose family is overweight is about five times more likely to have weight problems than the child whose parents are slim. So, do you inherit your body size? If your mother and father are both fat, do you stand any chance of being slim? There again, if your parents are both as thin as reeds, will you be like this too? No, you are not stuck with your body size. There is always something that can be done to change physical shape. Even if you inherit a tendency to fatness or thinness (and this is a much debated point), you need not be stuck with a weight problem. The correct food and the right type of exercises can work wonders for anybody. You can fulfil your own potential.

If a whole family has weight problems, it is nearly always the case that eating and exercise behaviours are wrong too. For example, members of a typical overweight family eat too much of the wrong kinds of food; too much fat, too much sugar. They do not have enough fibre in their food. Obese people generally have food piled onto their plates at mealtimes, and they tend to have a dessert after the main course as well. They never leave food. They snack a lot. They eat very quickly. They nibble at food when doing other tasks, or during leisure activities, such as watching television. They do not eat in just one room of the house but anywhere in the house, in the garden, in the car, or in the street. They eat if something looks nice, or if they have been offered food, or if it is a certain time of day. They eat when they are not genuinely hungry. Food is always available, in the kitchen of course, but also in the other main living rooms, in the bedrooms, in mum's handbag, and in the children's pockets. And, lastly, fat families tend not to be keen on exercise. Most of these things are unhelpful when it comes to controlling weight.

Slim families generally tend *not* to behave towards food in the same way as overweight families. If you want your family to be slim, it will help if you change to those habits

A quick look at family and friends

most likely to produce slimness. For example, it will help if you can do some of the following:

- Eat less sugar, fat, and salt.
- Eat more fibre. Have foods such as wholemeal bread, wholegrain cereals, peas and beans, fruit and vegetables.
- Do not pile plates high with foods, such as those high in sugar, or that are *low* in essential nutrients.
- Do not always have a dessert, or, if you do like to finish a meal with something sweet, serve plain natural yogurt mixed with fresh fruit, or maybe just the fruit itself.
- Leave food if appropriate. Do not eat once hunger has been satisfied.
- Do not snack. If you occasionally do feel genuinely hungry between meals, have a small quantity of nutritious food, such as a wholemeal bread sandwich, fruit, or a *few* nuts.
- Eat slowly, and savour the taste of your food.
- Do not eat mindlessly, such as when watching television.
- Just have one or two places in the house where you eat.
- Keep food in the kitchen and nowhere else.
- Introduce more activity into everyday life; for example, walk more, and walk faster.

Do not make all of these changes at once. If you go into your kitchen now and throw away all fatty or sugary things, take the sweets back from the children, ban eating outside of mealtimes, put a padlock on the food cupboard, cut down portions of food at meals, and refuse to give desserts, then *you will have a riot on your hands*! Do not go to extremes. You need to plan carefully the order in which you are going to introduce things. Remember, all this is supposed to make the family healthier and *happier* too! Setting appropriate goals for changing eating and exercise requires a little thought. Each family is different, so the goals set and the speed at which changes are introduced will be different for each family. In Chapter 7 we will be giving you tips on how to cope with your particular family.

It is possible, of course, that yours is a mixed family where weight control is concerned; that is, some of you are the right weight but some of you are not. Maybe just one or two people in your family have problems. A fairly common combination is the family with two children, one of whom is overweight and the other who is slim. Why should this be? There are a number of likely explanations, and perhaps one of these fits your family. For example, studies have shown that very often the fat child gets treated differently to the slim one as far as food is concerned. The tendency is for parents to think that the bigger child actually needs more food. The larger the child is the more food he or she is served at mealtimes. This, unfortunately, makes the child bigger still, and so the problem goes on.

Most parents are naturally very caring about what their children eat. Sometimes, however, good intentions can backfire. Over-encouragement and over-prompting to eat can lead to obesity. Under normal circumstances, a child will eat when hungry and stop when satisfied. However, this natural mechanism is going to break down if parents either force children to eat, or insist that plates are cleared of food at mealtimes. Also, if a youngster learns that eating is an easy way to please mum and dad, then there is a danger that the child will continue eating after hunger has gone, in order to get praise or at least to avoid being told off for leaving food.

Once a child has thus gained a reputation for being a 'good eater', then he or she has a reputation to live up to, overeating becomes a way of life. Internal cues of genuine hunger and of feeling full after eating get ignored. When this happens there is a very high risk of developing weight

A quick look at family and friends

problems. This will not happen with every child, not even within the same family. Some children, continue to listen to what their bodies tell them about hunger and feelings of fullness. Some children learn to please their parents in other ways besides overeating. But, generally, you can see that it is extremely important that you do not over-encourage children to eat. If they always struggle with food, then serve smaller portions. Do not offer desserts like cakes or puddings, which are high in fat and sugar. Do not offer snacks in between meals. If they are hungry, they can ask, and then have something small, such as fruit, to keep them going. Let your child have a chance to get back in tune with the internal signals which should regulate eating behaviour.

What if your child is too thin? Should you encourage eating then? Generally speaking, no. Some children learn that refusing to eat results in lots of attention and parental concern, which is really quite nice for them. You might have the best intentions, but at the same time you could be inadvertently encouraging and maintaining your child's faddy eating by being over-concerned. A good basic rule for any parent is not to fuss too much about eating. Make sure good nutritious food is available, of course, but do not be too pushy about the quantities that are eaten.

Naturally, children will have likes and dislikes for certain foods, the same as adults do. Luckily, no single food is absolutely essential for health. Very often it is possible to omit the thing that is disliked and replace it with something of a similar type. For example, if a child dislikes broad beans but likes peas, then serve peas. Maybe you can disguise certain foods. So, for instance, not every child will like fish, but if it is served finely flaked and mixed in with potato and a little grated cheese or a tomato, it may be much more acceptable. Experimentation is the name of the game. Even if a child is very rigid about likes and dislikes for certain foods, there is still a lot you can do to improve the diet: chips cut thick are less fatty than chips cut thin; just put a scrape of butter or margarine on bread; use less salt in cooking; instead of using tinned vegetables (which often have sugar and salt added) use fresh or frozen ones. The list of possibilities is endless. With a bit of imagination and cunning you can improve the diet over a period of time without anyone complaining that they do not like change.

The difficult child

It could be that you think that your family's diet is generally a healthy one, you encourage the children to play and exercise, but despite this you have one child with weight problems. Even if you do eat well, there are probably some improvements that can be made, and this will obviously be a help. But it is just possible that, despite your very best efforts, you find that your child is buying sweets, asking for extra helpings of pudding at school lunch, raiding the fridge, begging grandma for crisps and lemonade, and generally doing everything possible to sabotage your kind intentions to help with the weight problem.

In these circumstances you must be extra careful *not* to nag and *not* to punish—a miserable child is likely to compensate for feeling unloved by eating even more. Try to work out why your youngster is sneaking extra food. Is your child unhappy at school, or under some kind of stress? How is your child's social life? Are there lots of interesting things going on, or is your youngster a bit bored? If all seems well, then think again about what you are serving at mealtimes. A good, nutritious breakfast is necessary to start the day. This does not mean a big fry up. Wholegrain cereal, skimmed milk, fresh fruit, wholemeal toast are all good choices and will not involve you in too much preparation. If you serve good meals at sufficiently regular intervals, there should

A quick look at family and friends

not be any need to resort to sweet snacks throughout the day. High fibre foods are especially filling and nutritious, and should be available in some form at every meal. This is easily done *without* resorting to spoons of bran added to every dish. One mistake some parents make is to think that an overweight child should not eat bread, breakfast cereals, spaghetti, rice, potatoes, and starchy foods like this because they have something of a reputation as 'fattening foods'. Nothing could be further from the truth. In fact, they are high quality fibre-rich foods (when eaten in their wholegrain form: brown bread, brown rice, etc.) which are filling and hence prevent snacking on sweet or fatty things. If you give a child a very low calorie meal without much fibre, you will end up with a very hungry, miserable child who virtually *has* to resort to sneaking food or even bingeing. It is essential that anyone, adult or child, trying to control their weight should not feel deprived. A child who *does* feel this way will probably react against the enforced food regime at some point and have a real binge on all the forbidden foods. In no time at all everything will be back at square one.

If you want to help a child who is underweight, you actually need to do many of the same things that are necessary for helping a plump child. Make sure that the food you serve is nutritious and appetizing. Try to ensure that your child is happy, not under stress, has a good time with friends, has hobbies which are interesting, and so on. Do not fuss about eating. There is no need to over-encourage eating by leaving chocolates around, serving giant-sized meals, and frying everything you can at mealtimes! You want a child who is fit, not fat. Making your child add layers of body fat is not going to make for better overall health. Basically, if the food that is eaten is good, if your child is having appropriate exercise and keeping the muscles well toned, and if your youngster seems happy and well, then do not worry.

Parties and the like

Social occasions, where there are lots of special goodies, such as birthday cake, can be fraught with difficulty for the overweight child. (In fact, these occasions generally involve high fat and high sugar foods which are unhealthy for anyone, regardless of weight.) What should you, as the caring parent, do? It would be wrong to deprive your child of a good party, so should you pack your child a special bag of food to take along, instead of having what is on offer? No. Should you tell your child to have just one or two savoury sandwiches and refuse everything else? No. Should you instruct the adults supervising the party to let your child have only a little of the food on offer? No. It is much better for your youngster to feel just like all the others at the party. If you look at what children choose at parties you will find that not everybody likes and eats absolutely everything on offer. Your ultimate goal is to educate your child so that he or she can exercise a little self control. Many children and adults do this automatically, anyway. If we are going out to a big meal, we obviously want to enjoy it, so we make sure we are really hungry before we start. Maybe we eat a little less before the event. Then we learn that to appreciate good food and a variety of tastes, it is pointless gorging ourselves on a starter so that by the dessert time we are too full to cope. We can learn the art of appreciating good food only if we learn to space out eating so that food satisfies a true hunger. You will be teaching your child to enjoy food by eating in response to internal signals. At a party, ideal food for a healthy diet might not be available, but if your youngster eats just enough of any of the foods on offer to satisfy hunger, and then stops, the programme of weight control will not be disrupted. And should the food be so tempting that a little overeating does occur, then all is still not lost—if the child has eaten a lot, it should sustain him or her for longer. Do not insist that your child takes the next meal just because it is time to eat again. It is perfectly all right to skip the odd meal, or just have a very light snack, if you have overindulged and need to take a longer-than-usual break before proper hunger returns.

Another point to remember is that, for most children, parties do not occur that often. How many does your child go to in a year? Five? Ten? Twenty? An average of one or two a month would be quite high, but even so, consider how many other meals are eaten during the month—probably around 90. So a party meal is only a very small percentage of the food eaten overall, and therefore not too much to worry about. Part of the weight control programme that you will be establishing will involve your child being self motivated to achieve the correct weight and good physical shape. This motivation should keep your child on the right track to better health despite the occasional party.

School meals

These are something else! Some schools, to give them due credit, have tried to offer good meals. Most, however, seem to fall sadly short of nutritional perfection. Unlike parties, this can really be a problem, as your child could well be having school lunches week in, week out, for most of the year. If your child *has* to stay in school at lunchtime, or is keen to stay in to eat, there are several possibilities open. Packed lunches can be every bit as good as cooked meals, better in many cases. For example, banana sandwiches made with wholemeal bread, with maybe a few nuts and raisins, and a drink of low-fat milk is a lunch which is quite popular. At the same time, it is high in essential nutrients yet low in fat and refined sugar. If packed lunches are not possible, or your child would rather not take them, then you can enquire about the kinds of food on offer, and advise your child about appropriate choices. Also, as a health conscious, caring parent you could alway lobby for a better variety of food at school.

Presents

It is the custom for confectionery to be given occasionally as presents. Receiving a little special chocolate at Easter, Christmas, or on a birthday can be very pleasant for most people. Of course, other types of present are nice too, and it might be worth considering the possibility of giving other kinds of gifts besides things to eat or drink.

All this is very different, however, from giving inappropriate food as a token of affection on a daily basis. If neighbours or relatives give your children sweets, lemonade, crisps, ice-cream, or the like virtually every time they clap eyes on them, this spells nutritional trouble. It might be worth having a quiet word with them. Leaving the problem of weight control aside, sugary things eaten (or drunk) on many occasions throughout the day spell ruin to the teeth. If people want to show affection, there are other ways besides giving food laden with sugar or fat. If they really want to give children something to eat as a little extra to meals, fresh fruit is probably about the best thing. If they want to give something to drink, there is actually nothing wrong with plain water. Natural, unsweetened fruit juice is an alternative. Although this does not need to be diluted, it may taste a little acidic, and children often prefer it with water added; try fizzy water, such as 'Perrier', for a change. These suggestions apply to all children, regardless of weight. Even so, occasionally some well-meaning person is likely to give your child sweets or something similar. If this is a rare event, it should not be too much of a problem. There is no need to go purple with rage. Just make sure you are not one of the people who continue this rather unfortunate social custom.

Holidays and festive occasions

Do not forget to highlight the other activities that are going on which do not necessarily involve consuming food or drink! A trip to the seaside should mean more than fish and chips or candyfloss. Christmas should mean more than a turkey meal followed by mince pie or christmas pudding. People with weight problems can sometimes become so engrossed with food that they lose sight of what the occasion is really about.

But if you are planning the food for a special occasion, by all means stay with traditional meals. A turkey meal, for instance, can be highly nutritious. Poultry is a low-fat meat and, served with plenty of vegetables, this can be delicious and satisfying. There is no need to add a stack of salt to everything, or smear the vegetables with large pats of butter. Rich desserts like christmas pudding can be pleasant, provided they are not eaten to excess. The rule to remember about holidays and festivals is one we have already mentioned—do not eat unless hungry, and stop when you feel full. Children should be allowed to leave food if they want to. You also need to bear in mind that most special occasions are relatively brief. For example, it is all right to eat a little differently on the few days around Christmas. However, if you serve christmas cake as dessert at every meal for weeks after, this is not good news as far as good nutrition is concerned. Festive food should be confined to the festive occasion.

Summary

Since eating is often a social activity, it is important, when bringing about changes in eating, to guard against disrupting your child's social life. Similarly, since the social setting may influence eating patterns, you need to consider not just your child's behaviour but also your own and that of your family. As things are, you will probably find that eating occurs for a number of inappropriate reasons—the food is there, it is a mealtime, and so on. What you are aiming for is a situation where your child's main reason for eating is that he or she is hungry, and where he or she will stop eating once the hunger is gone. It is important that you do not make too much fuss about your child's eating, as this could lead to eating for the wrong reasons. Similarly, do not worry too much about the occasional bit of unhealthy eating at children's parties and the like. If your child's eating habits and lifestyle are generally healthy, coping with the once-in-a-while unhealthy meal should be no problem.

3 The parents' attitude to eating habits and exercise

Eating habits

It may come as something of a surprise to learn that the *parents'* attitudes to food and exercise might best be changed. 'Surely,' you might say, 'it is the *child* with weight control problems who could do with attitude change.' This is true too, but it is going to be far easier for your child to change if your attitude also alters. For example, many people regard giving food to someone as the direct equivalent of giving love and affection. If you offer your youngster a sweet or some other food and you get a reply of 'no thank you', what do you do and how do you feel? Do you simply accept the fact that your child does not want it, and think no more of it? Or do you feel slightly rejected yourself? You might even go on to say something like: 'Go on, just one won't hurt', or 'Go on, take it and have it later', or 'Go ahead, try it, it's really nice.' If you come out with any of these types of statement the child may feel obliged to take what is offered, not *necessarily* because the food is wanted, but because it becomes apparent that to refuse could make you feel hurt or rejected, or could cause offence. You should be sure that you are giving your child food for nourishment, not as an easy way of expressing love and affection.

With children it is often the case that they take what is offered because of the way you offer it. You almost force them to accept by saying something like 'Mummy has got something *really* nice here . . . I bought it especially for you . . . you'll love it I'm sure.' What child is likely to refuse? Of course, this does not matter if the thing you have bought is a *rare* treat. Trouble really starts, however, when this *pressing* offer of food occurs most days. You might not be presenting your child with sweets every time, it might be any food or drink. In the end, you do not even need to add persuasive words to the offer of food. You simply give your child lots of food and he or she understands that all this is a gift from you, a token of love, which if refused will lead you to be offended. Think about what you actually say if your child leaves food on the plate at mealtimes. Do you say 'Have you had enough? . . . OK, that's fine'? Or do you say something like 'Go on, can't you finish off that last little bit?' or 'What's the matter, doesn't it taste alright?' or 'You're not leaving that lovely food are you?' or 'I made that especially for you because you like it so much . . . you're not going to leave it are you?' All of these questions are likely to tempt the child to continue eating, in many cases after a feeling of fullness has already been reached. The child resumes eating not as a response to an internal signal of hunger, but in direct response to you, to please you. To refuse might cause offence, might make you feel rejected. Your child does not want to do this, and, therefore, overeats. It would make life a lot simpler if you did not press your child to eat up! Remember, you want your child to learn to eat in response to hunger, and to learn to stop eating when the hunger is gone.

Occasionally, the situation might arise (especially with a very thin child) where the child stops eating mid-meal, even though hunger is still there. The child wants you to press him or her to eat, as a reminder that you care. Generally speaking, you should do your utmost to make sure your child is happy and feels loved within the home environment.

The parents' attitude to eating habits and exercise

But make sure that your own attitude towards children's refusal of food is an appropriate one. You can see that meal times, the giving and receiving of food and drink, can have a great deal of psychological as well as nutritional significance! You should not feel offended if food is refused or left. Under normal circumstances this need not signify a refusal of love. Quite simply, it should mean they have had enough to eat! Throw the leftovers away, or, if they are suitable, use them again in another meal. Learning to change your attitude to other people's refusal of your food might be a really difficult thing for you, but it is absolutely essential if your child is to get in tune with internal signals of hunger and satiation. Normal children are not going to let themselves starve, so there is no need to fuss over food.

Of course, it might be that your youngster is so used to the idea that all food must be eaten, that food is never left on the plate anyway. You might need to make it explicit that leaving food is really acceptable. Do *you* ever leave food or refuse food? Children model themselves a lot on their parents. Try to stop eating as soon as you feel full. It will not *always* be when the plate has been cleared. Set a good example. Forget the idea that leaving food is a waste; eating *excess* food, which does more harm than good, is much more of a waste than leaving it on the plate.

Remember that even if you slightly modify the family's food, so that it is really a healthy, highly nutritious diet you are all eating, this will be of little use without appropriate attitude change. If you serve overlarge portions, and insist on everything being finished up (because it is 'good for you'), then weight problems are still likely to occur. Giving out fruit and nuts instead of sweets as between-meal snacks might be better for the teeth, but if these snacks are eaten just out of habit and not to satisfy true hunger (that is, because the body needs food), then this spells trouble as far as weight is concerned.

Exercise

Just what is your attitude to exercise? Is it the stuff that most children do until they are old enough to leave school and get a car—and from then on it is a slow decline to extremely inactive old age? Do you think activity is fine if you are young and fit, but dangerous if you are fat, or ill, or approaching middle age (for who knows when a heart attack might strike)?

Nowadays, there is more and more emphasis on the dangers of being *inactive* rather than active, at any age. The body is meant to move, to run up and down stairs, to walk about. As with your attitude to eating, it is important to realize that your children copy you to a large extent. If your attitude is that exercise is bad for you or not important, then this is hardly going to encourage the children to be active themselves.

Apart from playing with other children or taking part in sport, many youngsters are discouraged from taking part in physical exercise of any kind within the family. For example, many children simply do not have the opportunity to walk as much as they might, because even for short journeys (to the shops, to school, and so on) they are driven by car. Whatever happened to the family walk? Even when journeys *do* happen by foot, very often you see young children being pushed along in a buggy or a pushchair, despite the fact that they could probably walk the distance. At the very least, the child could be allowed to get out of the buggy and walk for a short while (attached to reins if need be).

There is no doubt that exercise is very important for general health and well being for all ages. It should form a part of any weight control programme. Exercise for children should be an integral part of everyday life, and, of course, it should be fun. Your child is not likely to regard it in this light if you consider it a total waste of time, or if you help your child avoid exercise. The trick, especially for a plump child, is to find an activity which involves a moderate amount of exercise but which, at the same time, is a real pleasure. For example, seek out pleasant countryside walks, take binoculars for bird spotting, or go fishing. Go kite flying. Find out what suits your child and what suits your family. Put as much energy into these activities as you can. Exercise need not mean competitive sport. Overweight children are not likely to do well at this kind of thing, and it may be understandable that they do not enjoy it. However, as physical shape improves, you will probably find that preferences change, and more activities and sports are tried and enjoyed.

Labelling your child

What exactly do we mean when we talk about 'attaching a label' to someone? As parents we are in a prime position to observe our children, to see how they behave, day in, day out. It could well be that we see our children tucking in and enjoying food—we might therefore use a label like 'a good eater'. Conversely, if a child always picks at food we might say this person is 'a poor eater'. It is a perfectly usual thing to draw inferences in this way.

However, having come to this decision, many parents (or indeed other relatives or friends) use this label a lot, probably a little too much. For example, if you are frequently remarking that your child is a good eater, has a really big appetite, really loves food, never refuses anything, then your child will come to believe that this is always the way to behave. You are greatly increasing the chances of the child overeating and becoming obese. There again, with the case of a thin child, constant remarks about being a poor eater will increase the chances of food refusal.

It is, therefore, very important that you do not continually go on about your child's eating habits. Even though you have all the best intentions in the world, you might be having a bad effect on your child's attitude towards food.

Flexibility

In terms of convenience, most families like to have certain meals together at particular times of day. Do not be too rigid about expecting your family to eat certain foods or exact quantities of food at meals. The time at which you eat might also need to vary, depending on other activities. If you are all out together, walking or cycling, shopping or having some kind of family outing, do not insist on being back for a certain time just because that is the time you eat. If your child has been extra energetic, has just swum 20 lengths of the pool or some such thing, a bigger meal might be required. So cater for this. You will need to be fairly flexible if you are trying to get your child (and indeed other members of the family) to be in tune with real hunger, real nutritional needs.

Illness

Occasionally we all go off our food, and this is especially true when illness strikes. Obviously, if your child is really ill then you will seek medical help. However, even with a relatively minor ailment such as a cold, it is easy to lose appetite. If a person is laid up and inactive for a while, appetite should naturally decrease because energy requirements are less than usual.

If your child does not want to eat much, because of illness, it is usually nothing to worry about; this phase soon passes. Just make sure that plenty of fluids are available and follow what your doctor tells you to do.

Your youngster might develop a fancy for certain foods during an illness. It is usually fine to pander to this, especially if the craving is for extra fruit, or for protein foods. Occasionally, however, your child might fancy very salty food or sweet things. What should your attitude be to this? With certain illnesses (those causing severe diarrhoea, for instance) it can be of benefit to have extra salt on food, and also to have some sweet things. If illness is a rare thing, and you wish to comply with special requests for certain foods from your child, then go ahead. However, giving children what is asked for is a very different thing to force feeding them!

Develop a positive attitude to change

How often do you hear someone say something like 'The poor child stands no chance of losing weight . . . just look at the rest of the family', or 'Our little one is no good at games', or 'Our youngest just can't slim . . . loves food too much'? All these statements show a negative attitude. What is more, such attitudes tend to be unrealistic. Any child, of any shape, in whatever family still stands an excellent chance of being healthy and obtaining an appropriate body weight. It does not matter if the only foods your child likes are fatty, or full of sugar or salt. Tastes can alter to *preference* for a healthier alternative. It does not matter if your child is hopeless at sport and hates physical activity. You will be able to find something your child genuinely enjoys and gets to be good at.

If you want your child's attitude to be positive then you must really develop a positive outlook as well. If your attitude is right, then this goes hand in glove with starting your child on the right track to a better shape and better health.

You need to be both positive and fairly adventurous in your approach. For example, if you take the negative attitude that there is no point in trying food with less salt because no one will eat it, then you are not even giving your child a chance to develop new tastes, are you? You must try out the suggestions we are making. Occasionally you may meet with some opposition—not every meal will be a raving success. Do not let it upset you. Do not make a fuss about it. Try something else instead next time.

Finally, do not let your family become obsessed with food, exercise, or weight, to the exclusion of everything else. Do not be forever commenting that this or that food is good because it has a lot of vitamins or fibre, or that this or that thing is bad because it is full of fat. By all means impart a *little useful* knowledge about nutrition along the way. We do this to some extent already, for instance most people know that sugar is bad for the teeth. But at mealtime you do not want to be giving a full run down of the nutritional content of every item of food on the plate. There should be other things to talk about!

Summary

We can see, then, that successful control of your child's weight may depend to a large degree on your own attitudes to eating and exercise. If giving your child food is the way you express your love and affection, this is not going to help your child to learn the habit of eating only when hungry. If you avoid exercise, your child will be quite likely to follow your example. In the same way, your attitude to your child's eating and exercise is important. If you label your child a 'good eater' or a 'poor eater', it should not be surprising when they act accordingly. What you want is a relaxed, realistic attitude to eating. Maintain flexibility at those

The parents' attitude to eating habits and exercise

times when your child needs to step outside good eating or exercise patterns, for example during illness. Do not make too much fuss about food, develop a positive attitude to change, and your child's attitudes will follow—leading to a better lifestyle and good weight control.

4 What is a healthy diet?

What should we be feeding our children to keep them at maximum health and at a good body weight? It can be surprising to look at lists of nutritional requirements for youngsters. Your child might only be half your height, but this does not mean that your child needs only half as much food as you do. Generally, we find that children, even those as young as seven or eight, often need as much food as their mothers. They need to fuel their growing, active bodies and so plenty of good food is required. As they get older and go into their teens, even larger quantities of food are needed. It is not unusual to find that teenagers eat more than their parents. And even very young pre-school children need relatively large quantities of food. We should not, however, be too concerned with just *quantity* of food. It is most important that we look at the *quality*, the types of food and drink that are consumed. Are you supplying your children with enough of the nutrients that they need to ensure good health? Too much junk food, high in fat and sugar and low in fibre, is likely to lead to overweight, or at least deficiency in some of the essential nutrients. If your child is underweight, you have to be especially careful to ensure that the food you are giving is rich in the things that children need for growth and good health.

So, rather than just worry about your child's weight, you should aim to get the *food* right. Once you have learned the basic steps towards a healthy eating regime for your child, weight control will not be too much of a problem. You must *not* get too obsessed with what your child weighs, popping your infant onto the scales morning, noon and night, worrying about fluctuations in weight from day to day. A once-a-week weigh-in is quite adequate. You will be much better employed by spending your time trying out interesting recipes and stocking the larder with the foods which will help promote maximum health.

Nutritional needs of children

Calories—what are they?

Food and drink supply calories, but just what is a calorie? And what have calories to do with gaining or losing weight?

A calorie is a measure of the *energy* supplied by food and drink. The food we consume, whether it is protein, fat or carbohydrate, provides us with fuel for our bodies, gives us energy. You have probably seen calorie books which list exactly how much energy value (that is, how many calories) may be supplied by different foods and drinks. Each person needs a certain amount of food just to keep the body ticking over. Extra is needed for general activities, the movement involved in daily life. And in children, a certain amount is needed to support growth.

In an ideal world, people would eat just the right amount of food to supply their total energy requirements, with none left over. For if extra *is* consumed, it is converted to body fat. You end up overweight. If, on the other hand, you do not take in enough calories, you draw on your own reserves of body fat and you lose weight. If you only eat when you are hungry, choose nutritious food of the right kinds, and do a moderate amount of exercise, you should not have too many weight problems.

How many calories do children need? Obviously, this is going to vary a great deal from child to child. If your youngster is tall and very sporting, then more calories are going to be needed than for a smaller, less active child of the same age. If your child is overweight and obviously too chubby, then generally the diet has been too high in calories. On the other hand, if your child is too thin, then the diet is too low in calories. Bearing these points in mind, here is an *approximate* guide to the number of calories children need to supply their daily fuel requirements. Between the ages of one and three years, they should have around 1300 calories (5500 kJ) a day. This gradually increases to about 1800 calories (7600 kJ) each day for seven-year-olds. By the time children are 10 or 12 years old, requirements go up to approximately 2000–25000 calories (8400–10 500 kJ) daily, with boys tending to need slightly more than girls. This is more than the 2000 calories or so that many adult women eat.

Just how many calories does *your* child take in each day? We do not want you to get too involved with calorie counting, but it can be quite interesting to write down everything your child eats and drinks for a day or so. Include each thing, the sugar sprinkled on the cornflakes, the gravy on the potatoes, the butter on the bread, and so on. Then get hold of one of the many calorie booklets that are widely available in shops and on magazine stalls (in the UK 'Slimming' magazine publishes a very full book of calorie values). Work out just how many calories your child is having. What many find find is that a large number of calories come from little snack foods and drinks throughout the day. It may not look much if your child eats a packet of crisps in the morning, an ice-cream or some chocolate in the afternoon, a few biscuits for supper, and has lemonade or squash to drink a couple of times in the day. But these are just the kinds of things which contribute many calories to the daily diet. For example, just these few things might add 1000 calories (4200 kJ). There can be more calories in a

What is a healthy diet?

glass of fizzy drink and two or three biscuits than in a full meal of fish, potatoes, and peas. Snack foods tend to be high in either refined sugar or saturated fat (or both). Sugar, as well as being ruinous to teeth, contains no useful nutrients, such as vitamins or minerals. Having too many of the daily calories supplied just by sugar and sugary foods is going to spell bad health for *any* child, and for many it means weight problems as well. Many of the snack foods like biscuits, chocolate, or cake are not only high in sugar, they are also high in fat. Nearly everyone, regardless of weight, would benefit from a reduction in fat. So, cutting down on these kinds of foods makes good nutritional sense. It will help the child who is overweight. It will also benefit the child who is underweight, provided, of course, that these foods are replaced with more nutritious alternatives which are low in saturated fat and sugar.

Some people become very involved with calorie counting. They weigh every tiny morsel of food, write down all the details, and try to be very precise in calculating the calorific value of food and drink. Some adults have controlled their weight very successfully using this method. If they restrict their calorie intake to, say, 1000 or 1500 calories (4200 or 6300 kJ) a day, then the excess weight disappears.

However, it can be very problematic to count the calories in your child's food on a long-term basis. For a start, it can be difficult to get accurate information from your youngster about little extras which may have been bought or consumed during the day. Then there is the problem of meals at school, or at gran's, or food eaten at a friend's house, and so on.

You might be able to get your child involved in the odd calorie counting 'spot check', maybe for a day or two each month, but more than this might be too hard. What is more, it is not particularly necessary or desirable for your child to know, in detail, the calorific values of all the food and drink that is available nowadays. A few simple guidelines are really all that is necessary: eat more fresh fruit rather than sweets, do not bother adding salt or butter to vegetables, have sugar-free drinks, and so on. Children should pick up these tips easily enough if they are taught in an easy, relaxed manner.

An important point for *you* to note, however, is that although we emphasize the type of food rather than the number of calories taken in, never attempt to make a *drastic* change in the number of calories your child has in a day. This could spell big trouble in terms of health. If your child is overweight, just eating slightly less than is needed each day will use up excess fat in time. Your child may speed the weight loss up by being more active too. Use your child's *weight* as a guide to the correct rate of progress, rather than the number of calories. There is no need to reduce calories to a strict 800 or 1000 a day as many adults do.

Also, if your child's problem is one of underweight, bear in mind that just a *tiny* increase in food intake will help—as little as 100–200 calories (420–840 kJ) a day more than are usually taken. The rule is never to make sudden, radical changes to overall calorie intake. Remember to concentrate on giving your child the right kinds of food, and encourage exercise and general activity rather than calorie counting.

Protein—where do you find it?

Protein is an important part of anybody's diet. This is especially true for children because protein is essential for the growth and repair of body cells. So, where do we find proteins? Nearly everyone has heard by now that meat, fish, dairy produce (milk, cheese, yoghurt), and eggs are all good sources of protein. However, this does *not* mean that in order to keep your family well fed you need to buy lots of expensive meats, or be serving massive omelettes all the time. It is important to remember that protein is also found

What is a healthy diet?

in many plant foods as well as in animal products. Useful quantities of protein can be obtained from cereal products (like bread, rice, spaghetti, breakfast cereals), nuts, seeds, and pulses (for example, peas, lentils, beans). If you eat a *mixture* of plant sources of protein, such as mixing a cereal product with a pulse product, this gives excellent quality protein. You do not need to eat meat to be well nourished. A good balanced diet does not need to be expensive. Examples of good meals which contain a mixture of plant proteins are: beans on toast; lentil soup and bread rolls; breakfast muesli; rice and peas; spaghetti and soya meat made into a bolognese (the soya bean and all its derivatives, such as soya meat and soya milk, are excellent sources of protein).

You can see that most people mix their sources of protein without even thinking about it. Not only do we mix our plant proteins, we tend to combine plant proteins with animal proteins, for example sandwiches (protein from the bread) filled with animal protein (such as egg, cheese, meat, or fish filling). Other examples are risotto, paella, pizza, rice pudding, macaroni cheese, breakfast cereals and milk. We generally tend to mix our foods in this way, and it is actually a very efficient method of getting all the protein we need.

In the past, books on diet and nutrition have often emphasized the importance of animal sources of protein and have not made much of the plant sources. Hence, many people may not realize that such basic foods as bread, rice, or beans are valuable sources too. In fact, it is likely to be healthier to have slightly less protein from animal sources but slightly more protein from plant sources in each meal. For example, you might be thinking of serving your child a cheese omelette (made from two or three eggs and 2 ounces of cheese). This meal is, in fact, very high in protein and other essential nutrients. However, it is also low in fibre (no plant foods) and rather high in fat. It would, therefore, be better to serve a few peas or beans with the meal, and perhaps include a slice of wholemeal bread. These plant sources of protein are high in fibre. Of course, these additions might make the meal a little too large for your child, so you could reduce the number of eggs by one. This automatically reduces the amount of fat in the meal, as eggs are quite high in fat. You could actually make the meal even less fatty by adding only half the original amount of cheese. Instead, the food could be accompanied by a drink of skimmed or semi-skimmed milk. This, like the cheese, is high in protein and calcium, but has the advantage of being lower in fat than cheese.

This is just one example of how to improve the daily diet. Basically, the thing to remember is that there are lots of advantages in consuming plant sources of protein. They contain more fibre than animal sources. This is especially true of pulses (peas and beans) and wholegrain cereal products (such as wholemeal bread, brown rice, and brown spaghetti). Adding plant products to meals will tend to give a wider variety of nutrients (such as vitamins and minerals). And, importantly, most plant sources of protein have relatively little saturated fat. Cutting down on fat is an important way to promote good health.

Minerals—such as iron and calcium

Minerals are the inorganic elements in our diet. The two minerals most people have heard most about are iron and calcium. These have very important functions in the body. For example, iron is necessary for the formation of red blood cells; calcium is important for good teeth and bones. We need minerals—iron, calcium and also others, such as phosphorus, magnesium, iodine, and zinc—in quite small amounts. However, for maximum health it is absolutely essential that we get all that our bodies require. This is especially true for growing children, whose need for some of

What is a healthy diet?

the minerals (such as calcium) may be just as great or even greater than for adults.

There is no need to be worried about all the various minerals and how much of any particular food you have to eat in order to get sufficient. You could drive yourself into a frenzy by looking up charts of recommended intakes of each and every mineral and then working out exact quantities of specific foods to supply your child's requirements of each. Do not become over-concerned. Minerals are present in many foods, such as the protein foods we mentioned earlier, and in many vegetables and fruits. For example, useful quantities of iron are contained in liver, kidney, beef, sardines, pilchards, shell fish, curry powder, beans (including baked beans), peas, lentils, nuts (especially almonds and Brazil nuts), porridge oats, wholemeal bread, soya products, dry figs, apricots, sultanas, and prunes. These are just some of the diverse sources of iron in our diet. Your child does not have to like and eat all of the foods in the list in order to get enough iron! For example, if your youngster hates meat, how about serving sardines on toast instead? Or, if your child loves cake, rather than giving shop-bought cakes (high in fat and sugar), make your own, healthier, high iron alternative. Use wholemeal flour, maybe a *little* soya flour, chopped nuts, sultanas, and other dried fruits. Experiment with different natural flavourings, such as mashed banana or grated apple. If you use enough fruit in the cake you will not need to add sugar. Do not use too much fat. Add a minimum of margarine which is high in polyunsaturated fat (such as 'Flora'). You can always give your child a slice of this kind of cake as a meal in itself, perhaps served with some fresh fruit and a glass of milk. A useful thing to note, by the way, is that iron, particularly from plant sources, such as oats, flour, or dried fruit, is much better absorbed by the body if the iron-rich food is eaten in the same meal as food rich in vitamin C (such as fresh fruit).

Before we leave the subject of minerals we would like to mention how important it is for your child to eat enough calcium-rich foods. All children need a great deal of calcium. The richest source of this material in most people's diets is likely to be dairy food. Milk, whether it is ordinary whole milk or low-fat milk (skimmed or semi-skimmed) is a very rich source of calcium. It also has the advantage of being relatively cheap! Giving your child a pint of milk each day really is a good idea. The low-fat varieties are especially useful for problems of overweight, or if you are trying to reduce overall fat intake. If your child cannot stomach milk as a drink you can always add it to food (such as mashed potato or home-made soups). Failing this, most children like plain yoghurt (unsweetened) if it is mixed well with fresh fruit. Yoghurt is especially good as a source of protein and calcium, and also has the advantage that it is low in fat. Cheese is also high in calcium and protein, but because it is so high in fat it is best to make a little go a long way—grating cheese always makes a small amount look like quite a generous portion.

Vitamins—get the food right and you do not need tablets

Vitamins can be found in many foods and drinks, such as the animal and plant sources of protein that we have already mentioned, in vegetables, and in fruit. Vitamins are necessary for health and the growth of body cells. The need for these substances is high in children, but they should be getting enough if you follow the advice on food that we have already given. There is no need to buy pots of vitamin tablets to supplement the diet. Some of these tablets contain extremely high doses, and although excess may usually be excreted by the body, this is not true of vitamins A and D. Large doses of these are poisonous. Your family will not

What is a healthy diet?

need to take extra vitamins if you are giving them the right kinds of food.

The major vitamins are A, the B family, C, and D. Vitamin A can be found, for example, in yellow-coloured fruits and vegetables (such as apricots, peaches, melon, carrots), and also in tomatoes, spinach, and peas. There are, of course, animal sources of vitamin A as well. Liver, kidney, dairy produce, eggs, and fatty fish, such as herring and sardines, are good sources. This vitamin is also added to margarine.

There is a whole family of B vitamins, but the major ones—B_1, B_2, and niacin—are found in plentiful supply in wholemeal cereals, wheatgerm, soya products, peanuts, liver, pork, and bacon. However, note that pork and bacon can be extremely fatty, and bacon is high in salt, so do not have these too often. Vitamin B_6 is found in cereals, eggs, meat (especially liver and kidney), and fish. B_{12} occurs in meat, milk, and eggs. This vitamin is unusual in that it cannot be obtained from any vegetable product, so 'vegans' (strict vegetarians who have no animal products at all) need to take B_{12} supplements.

Vitamin C is found in plentiful supply in fruits and vegetables.

Vitamin D is especially important for children as its presence is needed for the absorption of calcium. The body can manufacture its own vitamin D if exposed to sunlight. When skin is exposed to the sun this vitamin is formed. However, if yours is not an 'outdoor' family, you will have to be especially careful to serve foods which supply enough vitamin D. One of the best sources is fatty fish, the fish which tend to be reddish in colour rather than white (examples are herring, kipper, salmon, and sardines). Eggs are also a moderately good source of this vitamin, although they are unlikely to supply all your child's needs. Vitamin D is also added to margarine. However, do not spread vast quantities of margarine over food; although it has added vitamins, it does not supply minerals and proteins as do other vitamin-rich foods, such as fish.

Food for maximum health— the NACNE report

NACNE stands for the UK National Advisory Committee on Nutrition Education. The NACNE report made some strong recommendations about diet: basically that we should be eating less fat, less sugar, less salt, and more fibre. These points apply to everyone in this country, regardless of age. Everyone in your family (and it does not matter whether they are overweight, underweight, or just right) will be healthier and stand a much better chance of living to a ripe old age if they follow these recommendations. They are particularly important for children, for childhood is the time when the foundations for a healthy life are laid. We will take each of the points in turn and discuss them in some detail.

Fats

Generally speaking, we all tend to eat too much fat. However, having said that, we must point out that not all fats are the same. The real dangers are saturated fats—they make you fat and they may predispose you to heart attacks. There is a lot of this kind of fat in, for example, red meats and meat products (like sausages), butter, lard, hard margarines, ordinary whole milk, cheese, eggs, biscuits, cake, chocolate, crisps, pastry, and food most often cooked in saturated fat, such as chips. All these sources of fat are foods to be concerned about. Here are ways in which you can reduce the amount of saturated fat that your family eats:

1. Eat less red meat. Even very lean meat with no visible

What is a healthy diet?

fat still has fat in it. However, you can keep fat levels to a minimum by trimming off all visible fat. Avoid very fatty meats and meat products. There is a very high proportion of fat in many of these foods, such as chops, sausages, burgers, pork pies, and meat pies generally. Do not use the fat from meat to prepare gravy or sauces. In fact, do not use any kind of fat to prepare sauces. Occasionally have poultry (which is low in fat) instead of beef, pork, or lamb. Have plenty of fish meals too.

2. Use a minimum of fat in cooking, or on vegetables at the table, or on bread. If this is done, the lack of fat may not even be noticed. For example, with many kinds of moist sandwich filling, such as mashed banana or cottage cheese, it is difficult to tell if butter or margarine is there or not. With beans on toast, do not bother to butter the toast—the tomato sauce in the beans makes the toast moist enough. Experiment with fatless cooking, to find what you feel is an acceptable alternative. The use of spices and herbs (such as fresh mint instead of butter on new potatoes) can help during the weaning of your family away from fats.

3. Go easy with the high-fat dairy products. Have cheese in moderation. A total of three or four eggs per week is adequate for anyone. Choose low-fat dairy produce when you can—yoghurt and skimmed milk are both very good. For youngsters under five years some authorities argue that children are in danger of not getting enough calories unless they have full-fat milk. However, if your child is under five but is very chubby, it is clear that she or he is getting too many calories and a switch to low-fat milk would be one way of reducing calorie intake. In fact, there is a strong argument that says that we should be giving our children the opportunity to have a healthy, low-fat diet from as early an age as possible—why wait

until they are five? Certainly, a great many children under the age of five thrive and are extremely healthy on skimmed milk. The choice is yours. If you are a bit uncertain perhaps you would like to consider semi-skimmed milk, which has a bit more fat than skimmed milk, but less than ordinary milk.

As far as milk for young babies is concerned, the mother should breast-feed if she can, for as long as possible. If, for some reason, she does not want to breast-feed, or if she cannot, she should follow the advice of the baby clinic or health visitor concerning the special baby milks that are available. Do not give bottled 'door-step' milk to very young babies.

4. Avoid too many fatty snacks such as crisps, chocolate, and biscuits.

5. Salad creams and dressings such as mayonnaise are high in fat. Try low-fat alternatives, or use vinegar, lemon juice, or yoghurt to prepare alternative salad dressings.

6. Avoid excessive use of pastry in cooking. Instead, use potato as a topping in savoury products. If you eat shop-bought pies, leave some of the crust.

7. If your family is very fond of chips, cut the chips very thick as they absorb less fat that way. If you buy chips from a chip shop, it is best not to eat the very small, fatty pieces of chips often found at the bottom of the bag! If you buy pre-cut freezer chips to cook yourself, *do not* buy the crinkle-cut varieties—these absorb a lot more fat than ordinary chips.

These are seven tips to help your family cut down on saturated fat. It is nice to know, though, that not all fats are 'bad'. Some kinds of fat are absolutely necessary for good health. These are polyunsaturated fats and can be found, for example, in fish (especially the redder fishes like sardines, salmon, and kippers), wholegrain cereals, nuts, and seeds. Your whole family would do well to eat these foods. It is also worth mentioning that 'good' fats are found in some margarines (marked as high in polyunsaturates) and in some cooking oils (such as sunflower oil and safflower oil). A switch to these products is therefore a good idea. However, because all fats are so high in calories, use the barest minimum necessary in cooking and food preparation.

Sugar and other carbohydrates

In much the same way as there are 'good' and 'bad' fats, there are 'good' and 'bad' forms of carbohydrate. The 'bad' kind is sugar (sucrose), the stuff we buy in bags and that is added to so many of the foods and drinks available nowadays. This kind of sugar is a highly refined product which contains no vitamins, minerals, or protein—it just provides 'empty' calories. Generally, we all eat far too much sugar. It is difficult to avoid because it is added to a great many things, such as cake, biscuits, fizzy drinks and squashes, chocolate and sweets. It is even in some savoury products such as tinned peas and baked beans. Every effort should be made to avoid sugar. Here are some useful tips on how to do this. These suggestions apply to the whole family, irrespective of body weight.

1. Cut down on, or avoid totally, things such as biscuits, cakes, sweets, chocolates, jam, ice-cream, sweet drinks, fruits tinned in syrup, and desserts which contain sucrose (refined sugar). For the true sugar addict you will have to cut down sugar intake very gradually. For children you could perhaps have a special time of day for a sweet treat. Keep sugar-eating occasions to a minimum. Or, if you have the time, how about baking your own sugar-free biscuits, cakes, and desserts?

2. Have well-balanced regular meals with good protein foods, plenty of vegetables, and fruit. If the appetite is satisfied with savoury products, with good nutritious food, then a craving for sweet things is less likely to occur.

3. Snacks do not necessarily mean sweet things. How about a small savoury snack instead? Or, if the desire really is for something sweet, how about fresh bread with a small quantity of low-sugar jam, or fruit spread, or a mashed banana topping? A few nuts and some dried fruit also make a reasonable snack. The occasional fruit milkshake can also go down well with youngsters—simply put a piece of soft fruit, such as banana, into a blender with some skimmed milk. It is delicious even without added sugar.

These are all suggestions for cutting down on refined sugar, which, as we have said, is a kind of carbohydrate. But carbohydrate takes many forms, many of which are good for us. For example, it is fine to have carbohydrate as it naturally occurs (as starch) in foods such as cereals and cereal products, potatoes, and other plant foods. It is perfectly healthy to eat bread, breakfast cereals, rice, spaghetti (all wholegrain), and potatoes. This applies equally to people who are overweight. Carbohydrates may also occur *naturally* as simple sugars. For example, fruit sugar, or fructose as it is often called, and glucose occur in many fresh fruits. There is a sugar, called lactose, which is naturally present in milk. Generally speaking, naturally occurring forms of carbohydrate are desirable because the foods that contain them are also sources of protein, minerals, and vitamins. However, beware of packets of food or bottles of drink which have lots of sugar of any kind added.

Fibre

What is fibre? It is the part of the food we eat that is not absorbed by the gut; a type of complex carbohydrate found in the cell walls of plants. Fibre helps to keep the gut healthy, working smoothly and efficiently. People in the Western world tend not to eat enough fibre. The NACNE report recommended that we should eat more fibre. This will be especially helpful if you have a child with weight problems. However, do *not* buy bags of bran and start adding large spoonfuls of the stuff to your family's food. Bran, and indeed other forms of fibre, are readily available in many kinds of food; here are some examples:

1. Wholemeal breakfast cereals like 'Weetabix', 'Shredded Wheat', porridge oats, and muesli are all good for fibre. 'Weetabix' and most mueslis unfortunately do have sugar in them, but in relatively small amounts. You could always make your own muesli without sugar. You might even let your children choose their own favourite nuts or dried fruits to make up their own special muesli mix. They could then go on to choose one or two fresh fruits to chop up and add to the mix when ready to serve.

2. Wholemeal flour, wholemeal bread, and other brown forms of carbohydrate foods (like brown rice and brown spaghetti) all have a higher fibre content than their white counterparts.

3. Fresh fruit and vegetables are always good. All the pulses—peas, beans, lentils, sweetcorn—are excellent sources of fibre. It is probably best to have fresh or frozen vegetables because tinned varieties so often have sugar and salt added.

4. Dried fruit, such as figs, dates, apricots, and prunes are excellent sources of fibre.

You can see that the foods which contain fibre have already been recommended earlier in this chapter because they also contain very many useful nutrients. Serve at least one fibre-rich food at each meal.

Salt

While you are thinking about changes in diet, you might like to think about salt intake. Although salt intake does not relate in any simple way to weight, high levels of salt consumption have been linked to a number of health problems, such as high blood pressure.

There are many different kinds of salt, but the one we usually mean when we talk about salt is sodium chloride. We eat too much salt. We learn to develop a taste for it at a very early age. So, cut down on salt for the whole family. Gradually cut down the amount you use in cooking. It will not hurt if you use none at all. Even if a recipe tells you to add salt to a dish, do not do it. Salt is already present in a great many foods that we eat, so there is no need to buy it or add it to anything. If the family is in the habit of using the salt-cellar a lot at the table, a first step could be to fill it with a substitute such as potassium chloride salt, sold at health food shops and chemists (usually it is known by a brand name such as 'Ruthmol').

As well as avoiding salt itself, do not give your family foods with salt obviously added, such as crisps or salted nuts. Processed, smoked or cured meats, such as bacon, are high in salt and so should be strictly limited. Most soups are high in salt—read the labels to find out. Home-made soups (use lots of fresh vegetables and pulses) are far more nutritious; remember to leave out fat and salt.

What is a healthy pattern of eating?

There is not a simple rule to follow when it comes to

What is a healthy diet?

deciding on a pattern of eating. If your family has three square meals a day, that's fine. If you have more, smaller meals, that's all right. Generally, most children need to eat fairly frequently during the day. If you pack your children off to school with no breakfast it is hardly surprising if they buy sweets or biscuits in the mid-morning break. Try to plan things so that your children do not get hungry and tempted by food that really is not very good for them.

When you plan meals for your family remember the following points:

1. Choose a good carbohydrate food as a basis for the meal. Examples are: wholemeal bread, wholegrain breakfast cereal, any types of brown pasta or brown rice, or potatoes. Remember that such foods are good for protein, vitamins, and minerals, as well as fibre.

2. Pick some other protein foods as well. These could be plant sources of protein (such as nuts, peas, beans, lentils, soya) or animal protein—you might like lean meat or poultry, fish, egg or a little cheese, yoghurt or low-fat milk. Have any kind of mix of protein food that you fancy. Ring the changes and experiment a little. Use foods that are as fresh as possible and do not rely on heavily processed items.

3. You will probably want some sort of fresh fruit or vegetables with the meal. Of course, if you have already decided on, say, potatoes as your carbohydrate and peas as an additional source of plant protein, you will already have included vegetables. If you want to, add others and fruit as well.

4. Do you actually need, or want, to add fat to anything? Can you get away with not using any? Try to do this if possible. Otherwise, use a little polyunsaturated margarine or good quality cooking oil.

5. Remember to leave out the salt, or at least use a little less every time, until the family has become used to the new taste of things.

6. How about the little extras like herbs? Experiment with herbs and spices. But remember to steer clear of high-fat sauces and gravies if you can. Many of these products are also high in salt and so are best avoided or kept to a bare minimum. Read the labels on foods, and if they are high in sugar or salt try to do without.

7. Drinks at mealtimes (and between meals) should be free of added sugar. Water is perfectly fine for quenching thirst. Try giving fizzy water (like 'Perrier') for a change. Serving water in a nice glass with a few ice cubes makes it a bit more special. A drink of low-fat milk is perfectly acceptable as an accompaniment to food, especially if the meal does not contain other dairy foods, such as yoghurt or cheese. To occasionally have a good quality fruit juice (with no added sugar) is pleasant, although, generally, you get more fibre if you eat the whole fruit rather than extracted juice.

Ideas for meals

These are just a few, simple meals to give you a basic idea about healthy eating. Feel free to pursue your own ideas too, but try to follow the nutritional guidelines we have already laid out. All the foods we mention should be familiar to you. Our examples are easy to prepare and very practical. (If we decided to list exotic, difficult, or expensive recipes you might be less likely to follow our suggestions!) However *do* experiment with your own recipes to cater for your family's particular tastes. There are now many excellent recipe books available to give you ideas on low-fat meals, high-fibre meals, and so on.

What is a healthy diet?

Remember: where we mention cereal products (such as bread or rice) use wholegrain or wholemeal varieties if you can. If you want to add margarine or oil, use the tiniest amount possible. Whenever milk is listed, we suggest you use skimmed or semi-skimmed varieties. Try not to use salt at all, or sugar. If fruit is listed this means fresh fruit.

Breakfasts

Give as much fresh fruit as you like with any of these meals:

- Wholegrain cereal (such as 'Shredded Wheat', 'Weetabix', or sugar-free muesli) and milk
- Porridge oats (made with milk)
- Plain unsweetened yoghurt blended with banana. Add chopped dried fruit (such as figs or apricots) and chopped fresh fruit too
- Fish (kipper, sardines, or pilchards) with tomatoes on toast
- A little grated cheese, or an egg, with tomatoes on toast
- Beans on toast
- Breakfast in a hurry: milkshake made from milk, fresh fruit, a little wheatgerm or an egg.

Meat meals

Remember to keep portions of meat small. If you like, you can supplement the dishes with extra pulses (peas, beans, lentils), or soya products. Soya meat is a plant product of high nutritional quality. It can be added to most meat dishes (spaghetti bolognese, cottage pie, curry, stew, chilli con carne, etc.). Soya is a little bland if you eat it alone. However, if it is soaked in tomato juice rather than water, or if soya mince is added to a moist, spicy mixture such as a curry or a bolognese mix, it becomes very palatable.

Another advantage of using more plant products, such as

pulses and soya, and less meat is, of course, that meals are cheaper!

- Liver and onions, peas or beans, carrots, boiled or jacket potato
- Meat and vegetable casserole or stew
- Jacket potato, a small portion of any lean meat (especially poultry), baked beans or any other pulses
- Lasagne (soya mince or lentils make a good meat supplement) and salad
- Cottage pie, carrots, pulses or any green vegetables
- Spaghetti bolognese and salad
- Curry with rice.

Do not forget that if you are using pasta or rice, use wholegrain varieties. If your younger children find some of the meals too spicy, use only a little of the spicy food on the rice or pasta, and add a generous portion of plain yoghurt to dilute the taste.

Fish meals

- Fish fingers (grilled, not fried), peas or baked beans, boiled or jacket potato
- Fish pie (use white fish, tomatoes, peas, and additional vegetables of your own choice in the pie; top the whole lot with mashed potato and a little grated cheese)
- A portion of fish (not fried) with thick-cut chips and peas. If you have bought this meal from the chip shop keep the fat content down by leaving some of the batter and not eating the smaller, thinner chips so often found at the bottom of the bag!
- Sardines or pilchards on toast, with tomatoes
- Salmon, or prawn, and salad sandwiches

What is a healthy diet?

- Haddock, broccoli or spinach, peas, cheese sauce (made from skimmed milk with no fat added)
- Prawn curry with rice

Whenever you are making sauces, with any kind of meal, tinned tomatoes (no sugar or salt added) form a good base.

Egg or cheese meals

Both eggs and cheese are very high in saturated fat. Keep portions small and supplement with other foods such as skimmed milk, salads, vegetables, wholegrain products.

- Cheese, or egg, and tomato—either in sandwiches or on toast
- Jacket potato sprinkled with cheese, peas or baked beans
- Pizza (try making your own using a wholemeal flour base and just a little fat)
- Lentil or vegetable home-made soup topped with grated cheese
- One scrambled or boiled egg on toast
- Brown pasta shapes (shells, twists, or whatever) mixed with beans and sprinkled with cheese
- Cheese and potato pie, green leafy vegetables or carrots.

Serve fresh fruit and skimmed milk with *any* of the main meals that we have listed.

Hopefully, you can now see that healthy eating need not necessarily mean expensive food and complicated recipes. If you do not have much time, there is no need to rely on the heavily processed, ready prepared foods that are so often high in additives. It does not take long to make a sandwich, or grill a piece of fish, or boil an egg! You will need to adapt our recommendations to fit in with your own way of life, but making healthy, nutritious meals *can* be a part of a busy parent's routine.

Summary

If your child's weight is going to come under satisfactory control, you will need to establish healthy eating patterns. Good eating habits provide for all the body's needs—including calories (fuel), protein for the growth and repair of body cells (and sometimes for fuel), vitamins, and minerals. Detailed guidelines on healthy eating can be found in the NACNE report. This points out that we could benefit from substantially reducing the fat, salt, and sugar in our diets, and increasing the fibre.

In general, good meals can be planned around foods with a high carbohydrate content (also, these will often provide some protein). To this can be added additional protein from some other source, together with fresh fruit and vegetables. Try to avoid, or at least to minimize, fat and salt. Do not give drinks which have sugar added. The examples we have given show that meals prepared on this basis can be healthy, tasty, and nutritious.

5 The importance of exercise

Control of our weight is in theory quite a simple business, a matter of balancing the amount of calories we take in and the number we use up. In practice, one of the things this implies is that we can influence our weight not only by watching the calories we eat, but also by paying attention to the ways in which we use up those calories. Of course, the main way we use up calories is in physical activity, or exercise. Successful weight control can be helped considerably by paying careful attention to exercise levels. We know this both from our own observations and from scientific studies. Perhaps the most dramatic illustration comes when our televisions show us athletes lining up on the track for Olympic finals. One of the most obvious characteristics of such individuals is their excellent physical condition. Successful athletes are never overweight. Of course, we are not asking you to put your child through the rigorous and demanding training procedures used by people such as Daley Thompson; that much training is only required if you are aiming to become a world record holder! Simply as a means of helping with your child's weight control, exercise does not need to be nearly as demanding as it is for an Olympic athlete!

Of course, seeing that top athletes are never overweight does not *prove* that it is their exercise that keeps them in such good shape; after all, it could be that all top athletes are on very strict diets (they are not!), or that of all the people who take up athletics only the ones who are in good shape anyway get to the top. That is to say, it could just be that the people who are overweight do not succeed, so that in the top events the only ones who are left are those who started out the right weight. Thus, simply pointing out that people who exercise a lot do not have weight problems does not *prove* that the exercise prevents or cures these problems. To know for certain, we have to look at the scientific evidence. There is considerable evidence that exercise is an important, often a vital, part of any weight programme. For example, in a study in the USA obese laboratory animals were given a much reduced calorie intake. The animals which were left inactive were found to show little weight loss by comparison with others which were regularly exercised. This shows that it is possible to *maintain* quite a high body weight, once your child is overweight, with relatively few calories, as long as activity levels remain low. Effective weight loss may require *both* a reduced calorie intake and an increased level of exercise. Looking at things another way, a researcher who *doubled* the calorie intake of a sample of college students was still able to keep their weight constant by increasing their level of exercise. What this shows is that taking regular exercise can compensate for even quite dramatically raised levels of calorie intake. Indeed, it has been found that heavier schoolchildren are not always those that eat more; often the lighter ones eat more, but are also more active. The heavier children may eat less but may have only half the level of activity of other children—and as a result become overweight.

How exercise influences weight

The relationship beteen exercise and weight is basically very simple. Physical activity uses up calories. If a child's physical activity uses up more calories than are actually being eaten, then fat from the body is used to make up the difference. Similarly, if more calories are eaten than are used up in activity, then the extra calories will be converted to fat and stored as a 'reserve', to be used when output exceeds intake. Thus, if a child has a high level of activity and a low calorie intake, fat will be *used* up. If there is a low level of activity and a high calorie intake, fat will be *stored* up. Therefore, to lose fat we must use more calories than we take in; to gain fat we must take in more than we use. If your child is overweight, this is a sure sign that the intake of calories is higher than the number being used up in physical activity.

In practice, there are a few complicating factors. Perhaps the most important is the fact that 'activity' is used here in a slightly different sense than the normal use of the word. For example, simply sitting at rest in a chair is still 'activity' in the sense that the body is still using up a few calories—in keeping the muscles in their position, keeping bodily functions such as the heart and digestive system going, and so on. Similarly, when children are growing this means that the body has to use up a large number of calories both in providing and assembling the new tissue. What this means, in practice, is that a child with a similar level of exercise to an adult may still need more calories than that adult in order to maintain weight. This, of course, helps to explain why there are so many overweight adults compared to the number of overweight children—since many adults eat even more than children, yet their bodies, unlike those of the children, are not using any of this food as part of the normal growth process. As a result, the food that children use to grow *up* may cause an adult to grow *out*.

Another complication is that while the relationship between activity and *fat* is fairly simple, the relationship between activity and *weight* is not quite the same. Although, all things being equal, increased exercise may result in reduced fat, certain types of exercise may have the effect of increasing the amount of muscle in the body. In fact, a fairly small amount of muscle may weigh as much as quite a large volume of fat. As a result, the person who exercises may find that the amount of fat in their body goes down, but that the amount of muscle goes up; the effect of this may be to leave their total weight much the same as it was to start with. Of course, this is not necessarily a bad thing; a body carrying too much fat is simply overweight, whereas a body with increased muscle may be stronger, healthier, and more athletic. Thus, in using exercise as part of a weight control programme, it is important to remember that checking progress by weighing may need to be supplemented by actually checking *appearance*, so that a lean, athletic body can be distinguished from the earlier fat body of the same weight.

In fact, the kinds of exercise most useful in weight control tend not to be the ones most likely to put on large quantities of muscle; this is particularly true of children, whose bodies are generally unable to build up very large muscles. Exercise of the right kind can make a crucial difference to any weight control programme, for adults or children. In helping to control *your* child's weight, therefore, it is useful to put some thought into how you might improve your child's general level of activity.

Types of exercise

Obviously, there are as many types of physical exercise as there are physical activities—running, walking, talking, doing the washing up, playing competitive sports, or anything else which involves moving a muscle. For simplicity, it is useful to consider exercise not in terms of what the person does, but rather in terms of the effect the exercise has on the person. Customarily, these effects are summarized as the three S's—suppleness, strength, and stamina. By and large, most forms of exercise will have some effect on all three of these, but exactly how much, and in what way, will vary considerably.

Suppleness tends to be increased dramatically by exercises such as gymnastics and ballet, which involve a lot of stretching to enable joints to become more flexible. Strength is most obviously increased by such exercises as weightlifting, directed specifically towards the building-up of powerful muscles. Stamina is increased directly by exercises such as jogging, where a steady level of activity is maintained for a period of several minutes. Obviously, an exercise which increases one of the S's may well have an effect on others; gymnastics and ballet, for example, will usually increase stamina, and to some extent strength, as well as suppleness. For the purposes of weight control, the types of exercise with which we are most concerned tend to be those which result in increased stamina. That is to say, the type of exercise which is of greatest value in weight control tends to be steady, fairly gentle exercise maintained for a sustained period.

Of course, not all forms of activity burn up calories at the same rate. As we mentioned earlier, simply sitting at rest will burn up a few calories; however, walking briskly or jogging may burn up over 20 times as many, or more. From this it can be seen that even a fairly light increase in the level of physical activity can be enough to make a dramatic difference to the calorie intake/output balance. To burn up one pound (approximately half a kilogram) of fat it is necessary to do without, or burn up around 3500 calories. This could be done in several ways; the calorie intake could be cut, the calorie expenditure could be increased (by exercise), or, ideally, the balance could be reduced by using both of these methods at the same time. Thus, to lose a pound of fat in a month, the individual has to make a change in lifestyle equal to reducing the balance by 100 or so calories per day. For a child, half of this or more may be obtained by simply being involved in physical activity equal to brisk walking for a quarter of an hour, or so. Playing with friends, running around, walking, and exploring may mean that hardly any reduction in food intake is necessary.

In practice we do not recommend trying to reduce a child's weight simply by increasing exercise, any more than we would recommend trying to reduce weight by simply reducing calories. For good health it is usually desirable both to eat sensibly and to take regular exercise. Moreover, slightly increasing exercise and slightly changing eating habits is likely to meet with considerably less resistance than trying to make a dramatic change in either one alone!

How children normally get exercise

At first glance it might seem odd to suggest that children might take more exercise—after all, surely children, are already so active that there is no *room* for any more? Most children certainly are fairly active by comparison with adults. For a start, most children at school are required to take part in physical education and games classes. In addition, many children will involve themselves in a lot of activity in playing with their friends—going for walks, exploring, playing games, and so on. Each of these is likely to burn up a considerable number of calories, and help to prevent the child from becoming too fat.

Unfortunately this does not always apply to children who have weight problems. For a start, many of the things children do during timetabled exercise at school may vary considerably in how much exercise they involve. The good, fit, keen netball player may spend a lot of time during a game running from one end of the court to the other, jumping, passing, and so on. The poor player, too fat to run properly or not strong enough to be a valuable team member, may never be on the receiving end of the ball, and may spend most of the game wandering listlessly up and down being ignored by the keen players.

Another problem stems from the fact that physical activity just is not as much fun for a child who is not the right weight. Running, jumping, climbing, and even walking can soon tire a child who is very weak, or very fat. As a result, such children soon learn to play games which are not physically demanding—playing with toys on the living room carpet instead of being out in the fresh air, or watching television instead of playing sports with friends. Indeed, it is worth remembering that many children who *are* keen on physical activities will not particularly want to play with those who are not very good—the ones who are over- or underweight. As a result, many children with weight problems find that their friends are not the children who play lots of active games, but are ones, like themselves, who like less strenuous pursuits. In the more sad cases, children with weight problems may find that they do not *have* many friends, but are left alone to play more solitary, and often more sedentary games. One of the effects of increasing your child's exercise, therefore, may be to produce an increase in popularity by increasing the number of games and pastimes that can be enjoyed.

Characteristics of suitable exercise

We mentioned earlier that different types of exercise vary in their effects, some being better for strength, some for suppleness, and some for stamina. This last group, the ones which relate to stamina, tend to be the ones which have the greatest influence on weight and on body fat. So what are the characteristics of suitable exercises for weight control?

The first characteristic is that the activity should be more or less sustained. That is to say, it should be an activity which continues for some reasonable amount of time—minutes rather than seconds. A few short sprints down the road may seem very tiring and produce an incredible breathlessness, but it will not do a lot for weight control. If you look closely at such an activity, you will realize that the amount of time actually spent doing the activity may only be a matter of seconds—half a dozen flat-out sprints of 50 yards or so may be exhausting, but the total time spent actually running will, for most children, be less than one minute, although with rests in between each one, the total time may be over five minutes. To have spent those five minutes gently jogging at a steady pace (on a soft surface such as grass, using good training shoes) would have burned off many more calories and probably been a lot less unpleasant! Brisk walking or cycling would do equally well.

Which brings us to the second characteristic of suitable exercise—its intensity. The best exercises for weight control do not need to be forced or intense. A lot of people have been brought up with the idea that exercise must hurt if it is going to do any good. *This is nonsense!* Pain is your body's way of telling you that something is wrong, and it is unwise to ignore pain, or to set a level of exercise that makes pain inevitable. As far as weight control is concerned, it is much better to do gentle exercise for a long period than hard exercise for a short one. This is good news of course, because it means that the chances of maintaining your child's motivation (and hence your child's level of exercise) are that much higher. If exercise is painful or unpleasant, the child will soon find reasons not to do it. If it is gentle and pleasant, your child may not even recognize that it *is* exercise. Having said that, there is of course a limit to just how gentle it can be—after all, lying in bed all day is fairly gentle and painless, but it is no use as a means of weight control! What you are looking for is activity which involves the movement of major muscles and which, while not actually leaving your child breathless, does involve rather heavier breathing than usual. The feeling your child should have is one of making a definite effort but not actually being distressed. Such exercise is much more likely to be maintained than exercise which is uncomfortable or painful.

Which brings us to the last desirable feature of exercise for weight control—its regularity. For exercise to be effective in weight control it must be regular. To exercise for ten minutes a day is both easier and more beneficial than doing a single one hour session per week. Ideally, exercise should become a normal, habitual feature of daily life, not something which is slotted in to an hour or so of the weekend. Of course, it need not be the same activity every day, as long as it gives about the right level of exercise. Indeed, for many children it is much more useful to have a range of activities; they are less likely to get bored, and hence more likely to maintain the regular pattern of exercise that we are trying to establish.

Of course, it is quite possible that an activity which is very good for weight control will also have other benefits. A ballet class, for example, will involve quite a lot of sustained activity, and will usually need to be regular. Besides giving the right sort of exercise for weight control, such a class may also be of considerable benefit in producing increased suppleness, and who knows, your child may eventually grow up into a future Margot Fonteyn or Rudolf Nureyev! (Although we are afraid that, if your child is already overweight, doing ballet with a class of trim, lithe children may be the last thing he or she wants to do.)

Other advantages of exercise

As we just mentioned, it is quite possible that exercise will produce benefits additional to that of helping with weight control. Exactly what these additional benefits are, depends to some extent on what the exercise is and how it is done. Depending on what is involved in the exercise, you may also find the following;

1. Increased strength and suppleness. Although the exercises we are recommending for weight control are those which primarily emphasize stamina, it is quite likely that such exercise will also benefit the other two S's as well. Jogging, for example, will not only provide increased stamina, but will also make the legs considerably stronger, will probably make muscles in the body and arms slightly stronger, and, if done in conjunction with appropriate warm-up and stretching exercises before and after, will also result in increased suppleness. Remember, though, that prolonged periods of running on hard surfaces is not good for young bones. Be sure your child has good running shoes, not a pair of tired old plimsolls. For some exercises you may wish to consider a fourth 'S'—skill. Most exercises will produce improvement in the actual technique involved. Regular swimming, for example, will produce some improvement in actual swimming technique; not up to competition standard, for which specific coaching will be necessary, but enough to make swimming noticeably easier for your child. In games which involve a high degree of skill and strategy, of course, the potential for improving skill is that much higher.
2. Increased confidence. The child who is physically inept, unable to play active games, and who tires easily is unlikely to be confident. Increased exercise means that your child can have experience of getting better at physical activities, be more willing to join in the games of other children, and more likely to do well in such games. All this can do wonders for increasing your child's

The importance of exercise

confidence. In addition, the kinds of improvements in skill we mentioned above, together with a generally improved appearance, can mean that other youngsters will be keener to play with your child. The result can be an increase in both physical and social confidence which will help your child considerably.

3. Improved family relationships. One of the best ways of making exercise more enjoyable for your children is to make it part of an activity for the whole family. This can mean that for a specific part of the day the family as a whole is doing things together. Shared activity is a great way to strengthen relationships. Even if it does not involve the *whole* family, there will still probably be some benefit for those who are exercising together.
4. Easier dieting. Increasing exercise means that calories (food!) need not be restricted by as much (if at all) to reduce fat. If your child is trying to lose weight simply by dieting, you need to be extra careful to make sure that the food which *is* eaten contains all the essential nutrients. The biggest demand of the type of exercise that we are recommending is for additional calories—the need for other nutrients (e.g. protein, minerals, vitamins) will only go up a little. The result of this is that your child will be able to have more food, because of the calories being burned up by the exercise. This additional food will, on a sensible diet, also provide many of the other nutrients the child needs. If your child's diet has to be kept to a very small number of calories, this means an enormous amount of planning to ensure that the small quantity of food involved still includes all the necessary vitamins, minerals, proteins, etc. It is so much easier to give your child sufficient exercise to permit a food intake which will allow these things to look after themselves!

Summary

To summarize, let us just take a brief look at what we have said in this chapter. We have seen how changes in your child's weight are determined by the balance between calories taken in and calories used up. A child who takes in more calories than are used in activity and growing will become fatter; one who takes in less will become thinner. This means that exercise, as well as diet, can be used as a means of weight control. In practice, good weight control will involve both diet and exercise. The kind of exercise which has most benefit in a weight control programme is that which is fairly gentle and can be sustained for a reasonably long period; to be effective it has to be regular. Such regular, gentle exercise will allow your child to attain a satisfactory weight without drastic dieting. Done correctly, it will also be fun!

PART 2

The weight control programme

6 It is never too early to start

Babies—the first six months

What *is* it that it is 'never too early to start'? 'Surely', you are thinking, 'you can't mean weight control for very young babies?' Well, it depends on what is meant by 'weight control'. Having read this far into the book you should realize that weight control does *not* mean resorting to severe food restriction or faddy diets. This is out for adults as well as for children and babies. Instead, weight control should primarily result from eating the right balance of the right food. This is true for youngsters too. It is never too early to start having a health giving, nutritious eating plan.

So what is a good eating plan for the very young? For babies it could not be easier. Breast milk is absolutely perfect. There is no need to give any other drinks of any kind, no baby glucose drinks, no special fruit juices, nothing. Breast milk contains the right balance of the essential nutrients your baby requires. It also contains substances which help fight infection. It is better than any special baby milk you buy in tins or bottles. It has other advantages too. Your baby will be less likely to bring milk back up after feeds. And if this does happen a little bit, breast milk smells much more pleasant than regurgitated bottled milk. When your baby has a bowel movement, the contents of the nappy should be far less smelly if your child is breast-fed. It is also worth mentioning that breast-feeding helps in getting the mother's womb back to its normal size very quickly and effectively after the birth. Many women also find that the extra nutritional demands that producing milk places on them helps in the battle to lose any extra post-pregnancy weight, and helps them regain their figure. This does not happen overnight, but if you decide to breast feed exclusively for at least four months (preferably five or six months) this is likely to use up any surplus body fat you may have stored.

So, let us say you decide to breast-feed. It is important to let your baby suckle whenever the need arises (usually whenever you hear crying!). There is no necessity to restrict the time your child spends sucking. Your baby will stop feeding when full. Many mothers say they worry about not having enough milk to keep their little one satisfied. However, this should not be a real concern because the more a baby sucks at the breast the more breast milk is produced. If the demand increases, the supply increases. The converse is also true. If the demand decreases the supply decreases. It is important, therefore, *not* to doubt your own capability to produce milk, and not to give the odd bottle of baby milk (usually a cow's milk or soya based preparation). As soon as you start doing this your own supply will indeed dwindle.

Some women feel that their baby is not satisfied because feed times are so frequent—perhaps every 3–4 hours, day and night. Actually, this is not unusual. The pattern of feeding might even be quite erratic, varying from as little as one hour between feeds to as much as five or six hours. There may even come a time (often when the baby is around 6–9 weeks old) when you feel that you simply are not supplying enough milk because your child cries so often. Maybe your baby wants to feed every 30 minutes or so, through most of the evening. Do not worry about it. Go with this demand because your milk supply will adjust if you give it time. It could well be that this 'stoking up' with milk (often through the late afternoon and evening) is a prelude

It is never too early to start

to that lovely time when your baby starts to sleep through the night.

By the time your child is 8–12 weeks old you will probably find that you need to breast feed less frequently. The composition of the breast milk changes, gets richer, so that it is enough to satisfy your growing infant without giving other kinds of milk or introducing solids. In fact, many doctors believe that breast milk is the perfect food for the entire first six months of life. Nothing else is necessary, although just as an insurance vitamin drops may sometimes be prescribed for the baby. However, you will probably find that by the age of four or five months your child will start to become interested in food (special baby preparations) if you decide to introduce it. In fact, many clinics encourage you to begin your baby on solids (actually fairly 'mushy' semi-liquid preparations) at about 12–16 weeks of age. Remember, though, that it is *wrong* to believe that solid food is more satisfying than milk for young babies. It can be too rich and make your baby ill; there is even the possibility of very young babies choking to death. So, do not forget, it is dangerous to introduce solid food (anything else besides milk) too soon.

What should you do if you cannot, or simply do not want to, breast feed? Having decided that bottle feeding is the thing for you there is still plenty you can do to keep your baby's diet nutritious. The good news is that never has the special baby milk been better or safer. If you follow the advice of your health visitor or clinic doctor, you will have no trouble in choosing an appropriate milk for your baby. These special baby milk preparations tend to come as powder for you to mix with water. It is absolutely essential that you follow carefully the instructions for making up the milk. If you make it too watery, there will not be enough nourishment in it. But if you go to the other extreme and use too much powder, this thick mixture will not supply enough water to quench your baby's thirst. This could lead to constipation, severe dehydration, and, in extreme cases, death. (This would only happen if you continued with these thick mixes for a period of time, and did not give other watery drinks.) However, provided you follow the manufacturer's instructions carefully, special baby milk is perfectly adequate. If you are worried about your child becoming thirsty in between milk feeds, you can always offer plain water (boiled water which has cooled). There is no need to give glucose drinks, fruit drinks, or anything else. Do not, under any circumstances, start adding sugar to anything you give your young baby. You do not want to start rotting the little teeth as soon as they begin to come through. Nor do you want your baby to develop a taste for over-sweet things at this young age; it is likely to lead to weight problems and bad health. Try your best to give your little one the best possible start to a healthy life.

Weaning—six months to one year

This is the period when the baby starts to have other food besides milk. It may take a while (from about the age of six months to the first birthday) before your child is having most of the things that the whole family eats. You will doubtless receive lots of advice from your baby clinic, and from relatives and friends as well. Some of the things you hear may seem a bit complicated or confusing, so here are a few hints which will help you to establish your baby on a really healthy eating regime.

First, introduce just a little solid food at meals, the odd teaspoonful or so should be enough to begin with. Something very simple, such as baby rice, is probably the safest thing. There are lots of baby foods available and they are marked with full instructions about preparation. However, do read through the list of ingredients and avoid anything with added sugar. Even avoid products marked 'low in sugar'. Some manufacturers do omit the common type of sugar (sucrose), but then add other sweeteners which are just as bad (like honey or glucose). Avoid these too. Go for more savoury products or very simple foods. For example, there are some kinds of tinned fruit (or jars of fruit) which actually do only contain fruit. Simple cereals like baby rice should not have sugar added.

You should not need to use these baby foods for very long before you start giving mashed or finely minced ordinary food. In fact, there is no need to give packets, tins, or jars of baby food at all if you are willing to be careful in selecting the right foods and spend some time preparing them yourself. If you leave the introduction of food until your child is six months old, then he or she will very quickly learn how to cope with the new diet. Sucking, chewing, and swallowing food will happen quite naturally at this age.

So what kinds of meals should you be giving your baby during the weaning period? Follow the same basic rules that apply to other children (and adults too). Leave out sugar from food and drink, and do not give anything with added refined sugar in it. Omit salt from cooking and do not add it at table. Keep fat levels low. You can do this in several ways. One obvious method is to give low-fat milk rather than the ordinary variety. If you are worried about the controversy over the best kind of milk for children under five years of age, then compromise and give semi-skimmed milk. Better still, even though you are introducing solid foods, continue to breast feed. Breast milk is the ideal drink to precede or follow a baby's meal. You could do this until your child is eight or nine months old, then you might wish to just give a breast feed first thing in the morning and last thing at night. It is up to you how much you want to cut down, but bear in mind that it is quite handy if your child can use a cup! By nine months of age your baby should be able to sip drinks from a plastic beaker or 'trainer cup'. Use water or milk for this rather than fruit juices.

Another good, yet easy, way to keep fat levels low is to use plain, unsweetened yoghurt in meals. It might taste a bit bland to you, but most babies love it. Introduce it at about six or seven months. Mash fruit with it. Cottage cheese and fruit is another good combination. Fruits should keep the fibre level reasonably high. Do *not* add bran to food.

Basically, your aim is to get your baby eating more or less the same as you. To begin with, food will have to be very finely mashed, then when your child gets the hang of eating, food can be given in larger pieces. The next paragraph gives some idea of a typical day's food.

There will probably be an early morning drink of milk, preferably breast milk. Then for breakfast you might give a simple baby cereal with milk. Well-cooked porridge is also quite popular. Do not add sugar. Experiment with different toppings, such as mashed banana, finely grated apple, or puréed apricots. Lunch could consist of a milk drink (again preferably breast milk), followed by plain yoghurt and mashed fruit, or (as your child progresses) simple sandwiches. Wholemeal bread should be all right, with a scrape of a good polyunsaturated margarine. Try out different sandwich fillings such as mashed fish (no bones), or a little grated cheese and chopped tomato. Extra fruit is fine with any meal. Even babies as young as six months will munch into a strawberry, peach (skin and stone removed, of course), or a slice of melon. For the afternoon drink give breast milk if possible, alternatively cow's milk. For the evening meal try to adapt a little of the family's food for the baby. Mashed vegetables (like potato, carrots, peas) are fine. Do not use fat, but use plain yoghurt or milk to get the mixture to the right texture. If you are having rice or pasta as the basis of a meal, give a small portion to your baby. Again, yoghurt, cottage cheese, or a little grated hard cheese mixes very well. Add a few peas or beans too. Wholemeal pasta shells and peas are good foods for your baby to practice at self-feeding; they are relatively easy for little fingers to handle. Give milk as the drink to accompany this meal (breast milk if possible). For supper, you've guessed it, it is milk again!

You will obviously need to experiment and introduce new things at the pace which is right for your particular child. Remember not to force feed. Portions need only be small. There are good recipe books available for children's meals which will give you more ideas if you get stuck (we list some interesting books at the end of Chapter 10). In planning a healthy eating pattern for your child, your aim is to offer three good nutritious meals each day. In between, milk or water is fine. Please do *not* be constantly giving your baby little extra things to nibble at. It is all too easy to get your child into the habit of a little chocolate bar when you go to the shops, a biscuit or two with a drink, and so on. This is not a kindness where your child's health is concerned, and is to be avoided.

The pre-school years

You have a great deal of influence over your child during this time. Your child's attitude to food will largely depend on your own attitude and behaviour towards food. Now is the ideal time to get good habits firmly established.

Your toddler now has language and is going to understand if you begin to talk about food and drink—you can agree that chocolate may taste nice but as it is bad for you (it rots the teeth; it can make you fat) it is best to have it only occasionally. Now is the time to make good food fun. Experiment with the presentation of food. Make fancy shapes out of vegetables or fruit. Get a cookbook which specializes in no-sugar, low-fat recipes and let your child join in with the easy bits of the cooking. Be patient with your child's attitude to new tastes. A food which is refused this month may be a firm favourite next month. Do not assume that something which is disliked today is off the menu for evermore.

Basically, the more fun you can make a healthy diet, the more your child will want to stick with it. Parents sometimes worry too much about what their children will and will not eat, and feel that they have to give in to the children's demands. You often hear statements like 'She'll eat nothing but chips', or 'He hardly touches a thing at meals so I've got to give him sweets and crisps or else he'd starve'. Healthy children do not starve. If you refuse to give snacks in between meals, they will eventually get round to eating more nutritious food. Your child will only *eat* chips at each meal if you *serve* chips at each meal. So what if you just stopped? Some youngsters (and two year olds seem especially good at this) throw a tantrum if they cannot get their own way. Do not let yourself be controlled by your own child. There may well come a time when you need to be firm and just say 'no'. If you are *consistent* with this, if when you say 'no' you mean 'no', the tantrums will stop. Do not, whatever you do, just give in sometimes (when your child is really kicking up a storm and making a real din). If you do, your youngster will soon learn that the only way to make you give in is to really 'go to town' and make a terrible scene. This is not what you want! You should remember to combine firmness with kindness. Make sure your child does not push you around. Remember, make the changes fun, and make sure they happen.

Summary

In general, the sooner you start establishing healthy eating patterns the better. For babies, the most straightforward way to provide a healthy diet is by breast feeding. If for some reason this is not possible, alternatives to breast milk should be prepared with care.

For older children, one of the most important things to do is to avoid establishing habits such as the frequent eating of sweets. This may involve the occasional struggle, with some children throwing tantrums when they are first refused the snacks to which they have become accustomed. Do not give in to this, but be firm and consistent. If you make the changes gradual, and make them fun, they will usually be possible without undue distress.

7 How to change to a healthy eating plan for the whole family

A great many people would like to eat a healthier diet. There is so much medical evidence that we could benefit from *less* fat, sugar and salt, and *more* fibre, that it is easy to be convinced that a change to healthier food is desirable. But how to change? How do you alter your family's eating habits permanently? It is absolutely essential that you take a long-term approach to changing overall diet. Gradually introduce better foods as you slowly phase out less desirable ones. This is the best way to help any member of your family to get rid of weight problems for good. This chapter describes some basic steps to guide you.

Step 1 Get to know what your family's eating habits are like now, before you start to introduce changes

It is a mistake simply to launch yourself into wholesale diet change. Not everything about your family's present eating habits can be bad. After all, there must be something positive about the food you eat if it keeps you alive and gets you from day to day.

One of the best things you can do is to get hold of a large notebook or diary. For the first week do not introduce any changes at all to the diet. Continue to buy the same foods

How to change to a healthy eating plan

and serve the same sorts of meals that you are accustomed to. Do not tell your child with weight problems to eat any differently from usual. Use your diary to write down, in as much detail as you can, what your family consumes (both food and drink). For example, if for Sunday lunch you serve meat, potatoes, carrots, and peas, write it in your book in this sort of detail:

Joint of lamb, roasted in lard
Boiled potatoes (butter added at table)
Tinned peas
Carrots (butter added at table)
Salt added at table plus salt used in cooking the potatoes and carrots
Gravy (with fat from the meat added as an extra)
Drink of tea (full-fat milk and sugar added) for mum and dad, squash for the children

There is no need to weigh the items, but be careful to note any additions like salt, gravy or sugar, and what kind of fat (if any) is used in cooking or added at the table.

As well as keeping a record of the family's meals, keep as full a record as you can of what your child consumes each day. Ask what was for school lunch, what snacks were eaten at school, home, or at a friend's house, and so on. Make it absolutely clear to your youngster that this information will not be used in any negative way—if your child is super-honest and admits that two or three bars of chocolate were eaten in one go, do not look aghast or say that this is very greedy. Such a reaction could make your child feel bad, not about eating the chocolate but about telling you. The result may be the realization that one way to avoid your reaction is to conveniently 'forget' a few items!

Try to keep this diary for a full week, but if this is more than you can manage, two or three typical days will have to do.

Here is an example of a typical day's food and drink for one child:

Breakfast (with family)	Cornflakes Milk (ordinary full-fat) Sugar added to cornflakes Toast (white bread) with generous spread of butter and jam
Mid-morning (child alone at this time)	Glass of orange squash One chocolate biscuit One apple
Lunch (with mum)	Cheese sandwiches made with white bread, spread generously with butter and thick slices of cheese Cup of tea with full-fat milk and sugar
Afternoon snacks (child alone at this time)	One ice-cream Glass of orange squash Two sweets
Dinner (with family)	Chips (crinkle cut and fried in lard) One egg fried in lard Tinned peas Generous quantity of salt added at table One peach

Now, take a quick look back to Chapter 4 if you need to remind yourself of what a healthy diet looks like. Make sure you know what *are* the desirable foods in our diet, and also those things which we all tend to eat too much of (fats, sugar, and salt). Get three coloured pencils and decide on a colour code for underlining things in your diary. We suggest you use green for underlining the really good foods (such as fresh fruit, vegetables which are fresh or frozen, potatoes, wholemeal bread, and wholegrain cereals). Use red for underlining the really unhealthy foods (saturated fats, sugar, salt, and foods high in these things—such as pies,

How to change to a healthy eating plan

crisps, biscuits, cakes, sweets, chocolate, ice-cream, most squashes and fizzy drinks). There will be some items that you will find it hard to label as really good or really bad. Use a different colour pencil (say orange) to underline those things you are unsure about.

In the example of a day's diet that we have just given, how would we underline the various items that are listed? Which items are clearly very good? Not a great number! We would probably underline just the apple and the peach in green. Clearly bad items (to underline in red) are the sugar, butter, jam, orange squash, biscuit, ice-cream, sweets, lard, and salt. The remaining things do not obviously deserve red or green, so we would use the orange pencil with these to indicate that some improvement is needed. For example, the cornflakes are not too bad a choice but a wholegrain cereal like porridge or sugar-free muesli or 'Shredded Wheat' would be better (less sugar, less salt, more fibre). The milk is all right, but it might be better to have a lower fat variety. All the bread used is white; wholemeal would be preferable. Grated cheese would be better than thick slices of cheese as less could be used, thus reducing fat intake (the calcium level could be maintained by adding a skimmed milk drink to the meal). The chips would be better as boiled or jacket potatoes, or maybe as very thick cut chips or potato slices fried in sunflower oil. The egg would be better boiled than fried. The tinned peas would probably have salt and sugar added, so fresh or frozen would be better.

This should give you some idea of how to take a critical yet constructive look at your own food record. Do not worry too much about underlining absolutely everything in just the right colour. Simply make the best decision you can based on what you have learned from Chapter 4. The purpose of this little exercise is to get you to think in detail about your family's diet, and also to give you a permanent record of your present eating habits. You will have a few pages of writing which also contain, we would guess, a fair number of red and orange underlinings but not so many green ones. One year from now it will be very instructive to write out a one week diary again. By then lots of food habits should be changed. You should see some real progress. Hopefully, nearly all the markings will be green, with just a smattering of orange and hardly any red!

Step 2   What is in your cupboard?

This is another little exercise in really getting to know your family's food habits. You may be raring to start a weight control programme with your child, but just hold on for a little while longer. We want you to take a really good look at the way things are now and understand exactly what needs to be changed.

Use your diary again to keep a record. However, do not bother recording meals. Instead, use it to list the main contents of your food cupboard and fridge. (Alternatively, make a list of your main weekly shopping.) Use your coloured pencils again to underline those items you consider healthy (green), those which are unhealthy (red), and those which do not easily fall into either category (orange). Alongside each food or drink underlined in red or orange, write down a better, healthier alternative you would like to replace it with. You might like to list several alternatives for some items. Keep your options as wide as possible.

Step 3   Reducing sugar

You are not going to do this all in one go. You must make a full list of goals, lots of little ways to reduce sugar intake. The list for your family will be diffrent from any other family's. This is because each family eats slightly differently. The meals you serve might be similar in some respects to

How to change to a healthy eating plan

other people's, but not in all. Your family's tastes will, in some respects, be unique. So, we are not going to give you a rigid list of goals to work through. We would like you to make your own list and pick out one new goal to tackle each week. The goals you set will be partly dependent on the way you eat now. Some people already eat less sugar than others and will, therefore, require fewer goals to reach a low sugar intake.

Here are some tips on how a typical family might make a list of goals. (Remember, no family is exactly like this so-called 'typical' example, so your own list will doubtless be a little different from this.) This family usually has quite a bit of sugar. The week's diary has revealed that sugar is taken mainly in the following forms:

1. Drinks. Sugar is spooned into tea or coffee for all the family. The children occasionally have weak sugary tea, but most often they have squash or fizzy drinks.

2. Sweets, chocolates, biscuits, or ice-cream are readily available and are eaten nearly every day. Usually the children have at least one bar of chocolate or packet of sweets each day. Biscuits are given at supper time, and with a drink in the afternoon. Ice-cream is not eaten so often, perhaps two or three times per week.

3. At least one meal a day is followed by a dessert which contains sugar, for example custard, trifle, rice pudding, fruit yoghurt with sugar added, tinned fruit in syrup, cake.

4. At breakfast either a sugar-coated cereal is served, or cereal with sugar 'hidden' in it (for example most mueslis). Sugar is also added to cereals.

5. Jam or honey is often used as a spread.

6. Many convenience foods are seen to reveal sugar when

the ingredients list on the packet is read. Sugar is often an item in tinned vegetables, soups, and sauces.

Most people, in trying their best to improve their diet, would launch into all these six areas simultaneously, attempting to cut out sugar almost entirely. Unfortunately, the old habits can creep back remarkably quickly. So go slowly. Should our typical family make six goals, one for each of the six problem areas, and aim for one goal a week for six weeks? That is to say, in week one all sugar in drinks and squashes would go, in week two all chocolates, sweets, biscuits and ice-cream and so on. This is not the way to do it. It would probably be too difficult for most people. You are attempting to change the habits of a lifetime, and the chances are most unlikely that six weeks is going to be long enough to do this.

We suggest that for our typical family the aim should be to take about a year to reduce sugar intake (and even this is fairly fast progress!). The aim should be to tackle a new goal every one or two weeks. (If a goal seems a bit hard, it might be better to keep with it for two weeks instead of one, before starting another goal.) So, a list of between 25 and 50 goals would be the sort of length to aim for. There follows a list of the goals to reduce sugar for our typical family.

Drinks
Goals
1. Cut down sugar in tea and coffee from two teaspoons to one and one-half teaspoons
2. Only one teaspoon of sugar in tea and coffee
3. Only half a teaspoon of sugar in tea and coffee
4. No sugar added to drinks
5. Buy only half the usual quantity of fizzy drinks and squash (buy unsweetened fruit juice as a replacement)

How to change to a healthy eating plan

6. No fizzy drinks or squash (buy unsweetened fruit juice, try mixing with 'Perrier' water or plain water).

Sweets, chocolates, biscuits, ice-cream
Goals
7. Serve same number of biscuits, but buy smaller, thinner ones
8. Only give one biscuit at supper
9. Try baking sugar-free biscuits (use, for example, apple juice to sweeten), stop buying biscuits
10. Fruit for supper instead of biscuits (or, if children are extra hungry, a small slice of wholemeal bread spread with mashed banana)
11. Ice-cream only twice a week
12. Ice-cream only once a week
13. Ice-cream as a special treat only (for example, on outings to the seaside)
14. Maximum of one bar of chocolate or one packet of sweets each day
15. Chocolate bar or packet of sweets to be eaten during a certain restricted time period, that is not 'nibbled at' throughout the day, which is far worse for the teeth!
16. Chocolate or sweets only on 5 days a week
17. Chocolate or sweets on 4 days a week
18. Chocolate or sweets on 3 days a week
19. Chocolate or sweets on 2 days a week, this level is not too bad.

Remember to replace the chocolate or sweets with other foods if the family is hungry—experiment with savoury snacks, fruit and nuts, exotic fresh fruit, banana sandwiches, home-made no-sugar, low-fat biscuits, or no-sugar, low-fat cake.

Desserts

Goals

20. Only one meal a day to be followed by sugary dessert
21. Experiment with plain yoghurt mixed with different fresh fruits; use this for dessert on two days
22. Halve the amount of sugar in home-made desserts
23. Use sugar-free cake or pudding at least twice a week (remember to continue the yoghurt plus fruit dessert at least twice a week too)
24. Serve plain fruit as a dessert occasionally
25. Buy fruit tinned in natural juice rather than syrup
26. Experiment with stewed fruit (dried fruit soaked and heated with no added sugar), or baked apple stuffed with dried fruits.

Breakfast cereals

Goals

27. Buy a sugar-free muesli; serve with chopped fruit (at least once per week)
28. Try porridge with either mashed banana or grated apple (at least once per week)
29. Stop buying sugar-coated cereal
30. Do not put out the sugar bowl with breakfast; no sugar to be added
31. Read cereal packets; avoid ones with high sugar content, at least on most days
32. Experiment with home-made muesli; let the family add their own nuts, stewed fruit or fresh fruit to the mix.

Jam and honey

Goals

33. Buy low-sugar jam

How to change to a healthy eating plan

34. Try fruit spread (no sugar) (you may need to go to a 'health food' store for this)
35. Consider other savoury products as spreads occasionally (e.g. unsalted peanut butter—just a scrape is needed; cottage cheese)
36. Use fruit spread or honey only sparingly, and probably no more than once per day.

Convenience foods
Goals
37. Use sauces and pickles (with sugar added) more sparingly
38. Use shop-bought soups only twice a week maximum
39. Use shop-bought soups only once a week
40. Experiment with home-made soups—no sugar, less salt
41. Stop buying all tinned vegetables with sugar added, get frozen or fresh vegetables instead—one possible exception is baked beans, these are so popular with children, high in protein, high in fibre, and low in fat that an occasional serving is all right
42. Buy fewer 'convenience' foods—packets or tins of food with sugar added; remember that 'fast' meals can still be healthy and easy; keep a list of easy-to-prepare meals (e.g. sardines on toast, boiled egg with bread 'soldiers', wholemeal pasta shells with peas, and a little grated cheese).

You can see that at the rate of one goal a week it will take 42 weeks before all the goals are achieved. There is no need to work systematically through the list from beginning to end. For example, this family might get to goal 2 (only one spoon of sugar in drinks) and then in week 3 try a totally different goal, for example goal 33 (the low-sugar jam). They find this easy, succeed with it, and in week 4 try something else that looks relatively simple, for example goal 7 (buying smaller, thinner biscuits). Obviously, with this system only a certain amount of 'jumping about' between goals makes sense. For instance, goals 16–19 are graded in difficulty (fewer chocolates or sweets as you progress through the goals). Attempt the easiest goals first. So, do not try goal 19 until goals 16, 17, and 18 have been successfully completed. Remember, that as you accomplish each goal you should carry on doing it even when you start a new goal. So, in the example above, when goals 1 and 2 were achieved (fewer sweet drinks—just one spoon of sugar added), this new behaviour was continued in week 3 when the low-sugar jam was tried. In week 4, the family were having one spoon of sugar in drinks *and* a low-sugar jam *and* thinner biscuits, and so on. By the end of week 42, assuming one goal is accomplished each week, the 42 goals will have been achieved and will *all* be in action.

You will note that the goals do not eliminate refined sugar entirely from the diet. So, even at the end of it all, chocolate or sweets can be consumed on two days each week (see goal 19). This should allow for the odd treat or party. However, it could well be that after a little while longer on a healthy diet your family will not be so particular about sweets and chocolate and you can just cut these out altogether.

Step 4 Reducing fat

As we explained in Chapter 4, the *saturated* fats in our diet are the ones to be reduced for better health. Be as careful and methodical in making your list of goals to cut down on fat as you were with your sugar goals.

Again we will give you some tips on making the right kind of goals. Our typical family, who were having too much sugar, were also having too much fat. The diary reveals that saturated fat is mostly taken in the following foods and drinks:

How to change to a healthy eating plan

1. Dairy products. Full-cream milk (silver top) is the only kind consumed. Butter or hard margarines are lavishly spread on bread or smeared on vegetables. Cheese and cheese spreads are used virtually every day.
2. Meat is eaten about four times a week. This is nearly always in the form of meat products for the children (sausages, pies, burgers). Chops are eaten occasionally (about once a week). Fat is not trimmed off meat.
3. Cooking fats and oils are used frequently for frying or baking foods (at least one meal each day). Lard or dripping is used for frying. Lard or hard margarine is used in pastry. If vegetable oil *is* occasionally used it is a blended variety rather than a polyunsaturated type (such as sunflower oil).
4. Crisps or nuts are eaten as snack foods three or four times per week.
5. Chocolate biscuits, cakes, and ice-cream are high in fat (as well as sugar) and are eaten frequently, nearly every day.

The list of goals to reduce fat may be as follows.

Dairy products
Goals
1. Pour the cream away from the top of the milk
2. Buy semi-skimmed milk to replace silver top
3. Skimmed milk to replace semi-skimmed
4. Buy only three-quarters of the usual amount of hard cheese (buy more plain yoghurt instead)
5. Use a margarine high in polyunsaturated fats (such as 'Flora') to replace butter in at least half the family's meals

6. Have polyunsaturated fat to replace butter all the time
7. Stop using any fat on fresh vegetables
8. Use no fat in sandwiches with a 'moist' filling such as mashed banana or cottage cheese

Meat
Goals
9. Buy low-fat sausages (now available in many shops and supermarkets) to replace the usual ones
10. In meals with meat, serve less meat than usual and more bread or vegetables (like baked beans or peas)
11. Occasionally, try burgers made from a vegetable mix rather than meat
12. Meat *pies* of any kind only once per week
13. In making meat dishes, use potato as a topping rather than pastry
14. Trim excess fat off meat
15. Have chicken once a week to replace a red meat meal
16. Increase the amount of fish eaten to twice a week at least (occasionally using fish to replace meat will make for a better diet as far as fat intake is concerned).

Cooking fats and oils
Goals
17. Buy sunflower oil (or any polyunsaturated kind) to use in cooking
18. Grill or bake food, rather than fry, whenever possible (e.g. fish fingers, low-fat sausages)
19. Try to reduce the amount of fat used in cooking by at least a quarter (e.g. in cake recipes, pastry)
20. Reduce the amount of fat used in cooking by at least half

How to change to a healthy eating plan

21. Experiment with no-fat recipes for cake (browse through specialist cookery books for ideas).

Crisps and nuts
Goals
22. Have crisps only three times per week
23. Crisps twice a week
24. Crisps once a week
25. Nuts are a very 'rich' food, high in calories and should be eaten only *sparingly* as a snack. They should not be easily available and 'nibbled at' for long periods of time (e.g. when watching television).

Chocolates, biscuits, cakes, and ice-cream

These foods are high in sugar as well as fat. Goals for reducing the intake of these foods have already appeared in the section on reducing sugar, so they need not be repeated here.

At the rate of one goal per week it will take 25 weeks to achieve them all and have a reduced fat intake. As with sugar, there is no need to work through the list from goal 1 to goal 25. There is room for a certain amount of 'jumping about'. When you make your own list of goals, try the things you find easiest, the ones best suited to your own family, first.

You will be aiming to reduce fat at the same time as you are reducing sugar. So, as well as your sugar goal for any one week, you will have a fat goal as well.

Step 5   Reducing salt

As well as the salt we add during cooking and at the table, there is a great deal of hidden salt in food. The typical family would probably be getting too much salt in the following forms:

1. Tinned and packet soups
2. Tinned vegetables and many so-called convenience foods (read the labels of the foods you normally buy)
3. Salty snacks, such as crisps and salted nuts
4. Salted meats, such as bacon and gammon
5. Some fish is salted (such as salted mackerel)
6. Salt in home cooking
7. Salt added to food at the table
8. Some take-away meals, especially Chinese food, can be very salty (if you feel very thirsty after a meal it is likely to have contained a great deal of salt).

You will see that we have already considered most of the above items. Many of them are high in fat or sugar and so have already appeared in earlier goal lists. By eating less fat and sugar you will already have cut down on shop-bought soups, many tinned and packet foods, crisps and meat. Any nuts you eat should be unsalted.

Some fish is salted but not all. Fish is such a good source of nutrients that you need not avoid it. Instead, vary the kinds of fish you serve your family.

Take-away meals can sometimes be a problem. Even if you do not add salt to the food, it can already contain salt. However, how often does your family have a take-away meal? If it is a rare event then do not worry about it. Most people cook at home for most of the time, so concentrate on this aspect of the diet. With salt intake, the only goals you will probably need to make are those concerned with salt added directly at the table and in cooking. Here are some ideas on how to make appropriate goals.

How to change to a healthy eating plan

Salt in cooking
Goals
1. Reduce the salt used in cooking to three-quarters of the usual amount
2. Bring the salt down to half the usual amount
3. Reduce the salt used to a quarter of the usual amount
4. Finally, have no salt in cooking.

Salt at the table
Goals
5. Buy a low-sodium salt or potassium chloride salt, such as 'Ruthmol'; mix ordinary salt with this new kind in the propotion: three-quarters ordinary salt to one-quarter low-sodium or 'Ruthmol', and fill the salt pot with this mixture
6. Fill the salt pot with half each of ordinary salt and 'Ruthmol' or low-sodium salt
7. Change the proportion in the salt pot to three-quarters of 'Ruthmol' or low-sodium salt and one-quarter ordinary salt
8. Just have 'Ruthmol' or low-sodium salt in the salt pot.

It can take quite a while to get used to a low-salt taste. There are only eight goals in this list and, therefore, it would not hurt to spend as long as one month on each goal. Alternatively, some people might prefer to tackle salt in cooking *and* salt at the table at the same time, and therefore tackle the goals like this:

> Goals 1 and 5; for two months
> Goals 2 and 6; for two months
> Goals 3 and 7; for two months
> Goals 4 and 8; for two months
> ———————————
> Total 8 months

The choice is yours. Do whatever you find easiest. You should be reducing the salt in your child's diet at the same time as you are reducing sugar and fat. Each week you will need to note in your diary which sugar goal, fat goal, and salt goal you are on that week. Your entire list of goals should be clearly written down together at the beginning of the diary. Remember to tick these things off as you go. It is important to keep track of where you are in the scheme of things!

Step 6   Increasing fibre

After all this advice on reducing things, it is lovely to find that you can actually increase some kinds of food. This applies to everyone, including people who are overweight. Foods high in fibre tend to be very filling, but at the same time not very high in calories. So, by eating plenty of fibre-rich foods you will be having a good quality, nutritious diet which will sustain you and make you much less likely to nibble at fatty, sugary snacks.

Luckily many of the foods with fibre are very popular, both with adults and with children. There should, therefore, be no great need to introduce these things slowly. A healthy diet should not be simply a list of 'don'ts'—things to miss out. People often fail because they do not consider the positive aspects of healthy eating. Concentrate on recipes and meals which contain the following kinds of things:

1. Bread preferably wholemeal (not just 'brown bread')
2. Wholegrains (such as wholegrain rice and pasta), wholegrain breakfast cereal, wholegrain flour in cooking
3. Peas, beans (all types), lentils
4. Fresh and dried fruit
5. Vegetables (do not overcook them), including potatoes

How to change to a healthy eating plan 99

(especially baked potatoes, so you can eat the skins as well).

Keep a list of interesting, easy-to-prepare recipes at the back of your diary.

Step 7   Getting under way

Once you have made your lists of goals you will be ready to get started properly. Write down your first goals (one for sugar, one for fat, one for salt reduction) at the beginning of the current week in your diary. Try your best to keep to them. Next week, write down your new goals, and so on. You have broken down the huge task of diet change into many easy peices which should be simple to do. Maybe you think your goals are too simple. However, if you can get through them all in one year, and can keep to them, this will be a great achievement. Your diary is there to remind you of where you are in the overall plan.

There is no need to write down every meal you eat (unless, of course, you want to). However, you might like to make a note of anything you find difficult, or any really helpful thing you have done, such as trying a new health food shop, or finding a super cook book.

Write down any food-related habits you change. For example, you might decide to stop telling your children to 'Finish up all your food'. You might stop offering food so often. Perhaps you might take meals at a more leisurely pace. You might decide to stop keeping food in any other room besides the kitchen. It might be that you need to ask relatives or friends to be a little less generous in handing out sweets. Note all these kinds of things in your diary.

Should you explain to your family what your plans and goals are all about? It all depends on the kind of family you have. If you think they will resent any change, then do not present them with your plan of action. Just get on with it. On the other hand, if they are keen to pursue a better diet for better health, then by all means explain the steps you are taking.

However, do not harp on endlessly about it. Especially, do not overemphasize all the things you are cutting down. You may find that you are presenting a very threatening situation if you only talk about reducing things. The emphasis should be on the new, healthier way of eating, interesting recipes, and so on. Your children should not feel deprived of anything, including sweet-tasting foods. If they are very fond of refined sugar products, make sure you replace them with healthier alternatives which are naturally sweet, such as fruit, baked apple desserts, home-made cake sweetened with dates, mashed bananas or grated apple. Experiment, then experiment again until you feel satisfied that you have found a popular replacement for your usual foods.

What will you do when things go wrong? Life seldom runs as smoothly as we plan. Do not expect to work systematically through each and every goal with no hitch at all. Expect the odd problem and try to be prepared for it. For example, if you really get stuck on one goal, leave it and return to it later. If you still get stuck, try breaking the goal down into smaller, easier mini-goals. Make sure that you are not being over-enthusiastic with your lists. A *little* saturated fat, or sugar, or salt may be necessary to keep everyone happy. Expect to make the odd mistake, the occasional recipe which tastes awful. You might also face opposition occasionally. For instance, your children might make it absolutely plain that the only kind of bread they like is white bread. What should you do? Remember, bread comes in all different kinds, all shapes and sizes, fresh and less fresh. Stale white bread can taste pretty awful. So can stale wholemeal bread. How about trying really fresh, slightly warm wholemeal bread spread with a favourite topping? Mashed banana is a good covering for bread. Use pieces of other fruit, such as pieces of peach and seedless grapes to make patterns on top of the banana. Use a bit of imagination to tempt the palate.

At the end of the year review your progress. Hopefully, all your goals will have been reached. Perhaps you might have added a few more to your original list, or dropped one or two which were just too difficult. Your diary should be full of comments about your progress, little tips and recipes which will help you maintain the new healthy way of living.

Summary

Bringing about a change to a healthy diet is not something which can be done overnight. Careful planning is necessary. You will need to know in some detail what your family's present eating habits are, keeping a diary in which you note healthy foods, unhealthy foods, and those which fall somewhere in between. Check what your current food stocks are like—do healthy or unhealthy foods predominate?

Once you know where you are starting from, you can start to introduce changes, reducing sugar, fat, and salt while increasing fibre. Be flexible, set goals carefully and work towards them gradually. Feel able to adapt your plans if you run into difficulty. Above all, do not try to rush things—you are setting up the habits of a lifetime, so it is worth going about it slowly and carefully.

8 Setting up a weight control programme with your child: diet

Funnily enough, this chapter is one which you may not need. You might think that if your child is underweight or overweight you should give them a special diet and the problem will soon be cleared up. Nothing could be further from the truth. A child *must* be seen in the context of the whole family. We have already shown you how to adopt a healthier diet for *all* the members of your family. We have encouraged you to take more exercise and indulge in more energetic activities with your children. We have also pointed out the importance of taking the right attitude to eating behaviours. So, for example, you should not force a child to finish off everything served at a meal—do not feel offended if food is either left or is refused in the first place. If you can go some way towards doing all these things, you might well find your child's weight control problem just disappears, slowly and gradually over a period of a year or so.

It might be tempting for you to look for an easy way. Say, for instance, that you have an overweight child to whom you start to serve nothing but calorie reduced, nutritious meals for a couple of months. Probably your youngster will get miserable, or feel left out of family meals. But even if this does not happen, and your child *does* actually get down to a good weight, then what? There is a very good chance that if your youngster just goes back to the kind of eating that caused weight problems in the first place, it will not be too long before the scales start registering a big increase in fatty tissue! Then it is back on another diet again. And so on, and so on, and so on.

This awful state of affairs should never be allowed to happen with children. It does their health no good and psychologically it is bad too. They never feel that they have really achieved anything.

So, do you really need a special weight control programme just for your child? As we have said, you may not. Start serving better food at mealtimes and all may be well. However, it can be fun to set up a specific programme just for your child. This will not be a boring diet sheet. It will be something like a contract. The contract will contain certain goals that your child would like to aim for (weight goals; foods to eat more of; foods to eat less of, and so on). These are the things your youngster should try to achieve. Your part of the contract will be to promise a reward for each goal reached.

This kind of weight control programme can add real incentive for eating better food, doing more exercise, and generally adopting a healthier style of life. It has many advantages. For example, because it is a special contract between you and your child, it can be geared to particular problems that your child has. It might be, for instance, that your youngster is the only one in your family who is a tomato ketchup fanatic. You know the kind of person we mean—someone who smothers every meal except the breakfast cornflakes with about half a bottle of red, sticky sauce. If this is a problem that only your child has, then you might find it appropriate as a goal in the contract. We will show you how to bring this kind of problem into the contract later in the chapter.

A personal weight control programme which you design

for your child can take account of all the funny little eating habits that only he or she might have. If you have more than one child with a weight problem, you can have a different contract for each one.

How to design a weight control programme for your child

You should draw up a simple, easy-to-understand contract between you and your child. The contract will consist of the following.

Goals

1. Appropriate goals for weight
2. Some goals for diet change
3. Goals for eating-related behaviours (such as learning to eat more slowly, leave food when full, and so on).

These things are topics for discussion between you and your child. Do not just present a list of hard goals. Sit down together and work out goals which are attainable. Go for lots of small, simple goals rather than a few great big ones.

Just as important as goals are the incentives you give for attaining goals.

Rewards

1. Give appropriate rewards for keeping to target weights
2. Rewards for diet change are important
3. Do not forget the rewards for good eating-related behaviour.

Again you will need to sit down with your youngster and work out a list of rewards. What does your child value? What can you afford to give? What do you both consider to be a fair reward for each goal? You need to get down on paper a list of all the goals and next to each one the reward you both agree on.

That, basically, is what the programme is all about. Now here are some tips to help you on your way.

Deciding on goals for weight change

What should children weigh? It is impossible to give an absolutely perfect weight for your particular child. There are too many factors to be taken into account. All we can do is suggest a range of average weights for children of a certain age and a certain height. If your youngster is not an average height for his or her age, then use *height* rather than age as a guide to weight in the tables.

Remember, though, that the tables of weights can only be approximate. By far the best indicator of whether your child has a weight problem is simply to look objectively at your youngster when he or she is undressed. Compare your child with others of the same height or age when they are playing on the beach or in the swimming pool. If you really cannot decide if your child is a little overweight, or underweight, and if nobody else thinks there is a problem, you are probably worrying unnecessarily. If, on the other hand, your child appears to weigh twice as much as any classmates, and you have difficulty buying clothes to fit, there is obviously a real problem. You will not need to look up weights in a chart to realize that something is amiss!

You will notice from the tables that the weight range for any particular age is really quite great. So, for example, ten-year-old girls tend to range in weight from about 4st 2lb (26.5 kg) at the smallest, to 6st 4lb (40.0 kg) at the largest. The chances are that if you have a girl of ten years who is of average height and build, she should be somewhere *between*

Table 1 Weight guide for girls

Years of age	Height		Average weight range	
5	3' 6½"	108 cm	2st 9lb–3st 5lb	17.8–21.5 kg
6	3' 9½"	115 cm	2st 13lb–3st 10lb	18.5–23.5 kg
7	3' 11½"	121 cm	3st 3lb–3st 12lb	20.5–24.5 kg
8	4' 2"	126 cm	3st 8lb–4st 11lb	22.5–30.5 kg
9	4' 2½"	128 cm	3st 10lb–5st 6lb	23.5–34.5 kg
10	4' 5"	134 cm	4st 2lb–6st 4lb	26.5–40.0 kg
11	4' 8"	142 cm	4st 6lb–7st 1lb	28.0–45.0 kg
12	4' 10½"	148 cm	4st 10lb–7st 11lb	30.0–49.5 kg
13	5' 1"	155 cm	5st 10lb–8st 10lb	36.5–55.5 kg
14	5' 2"	159 cm	6st 6lb–9st 5lb	41.0–59.5 kg

Table 2 Weight guide for boys

Years of age	Height		Average weight range	
5	3' 7"	110 cm	2st 9lb–3st 7lb	17.8–22.0 kg
6	3' 9½"	115 cm	2st 12lb–3st 13lb	18.0–25.0 kg
7	4' 0"	122 cm	3st 4lb–4st 7lb	21.0–28.5 kg
8	4' 2"	127 cm	3st 8lb–5st 3lb	22.5–33.0 kg
9	4' 4½"	133 cm	4st 1lb–5st 1lb	26.0–36.5 kg
10	4' 6½"	138 cm	4st 5lb–6st 4lb	27.5–40.0 kg
11	4' 8½"	143 cm	4st 7lb–6st 12lb	28.5–43.5 kg
12	4' 10"	147 cm	5st 1lb–7st 8lb	32.0–48.0 kg
13	5' 0"	153 cm	5st 7lb–8st 7lb	34.0–54.0 kg
14	5' 3"	160 cm	6st 2lb–9st 9lb	39.0–61.0 kg

these two weights. This, of course, still leaves you with a lot of scope! How *do* you decide on a goal weight for your youngster?

The simple answer is that you do not give an absolute goal weight. You cannot say that your child should be exactly so many stones, pounds, and ounces (or kilogrammes).

You *can* say that if a child of a certain height is outside the range of weights given in the table, this is likely to indicate a weight problem. However, if your youngster's weight is within the range, there might still be a size difficulty. With our example of a ten-year-old girl, she might really need to be somewhere towards the bottom end of the range, around 5st (31.5 kg). If she actually weighs 6st 4lb (40 kg), the top of the range, she is about 18 lb (8 kg) overweight. These tables cannot give you that kind of detail. However, this amount of excess fat should be obvious on a girl of this height. Rely on physical appearance as your first measure of appropriate size.

Let us go back to our original problem, how to decide on a goal weight. First, let us take a child who is overweight. Use the weight tables plus a good objective look at your child to assess the extent of the problem. Decide whether your child is:

- Just a tiny bit plump, probably no more than 10 lb (4.5 kg) or so overweight
- Moderately overweight, about 10–20 lb (4.5–9 kg) too heavy
- Frankly very plump, at least 20 lb (9 kg) overweight.

The goal weights that you set will depend on the size of the problem.

If you think your child falls into the first category, and is just a little too plump, there is no need for any quick action. The goal should simply be to not *add* any extra weight for a while. This is what you reward. As your youngster is growing taller, he or she should automatically appear slimmer as time goes on. Have a weekly weigh-in and give a weekly reward for not adding weight. Make it clear to your child that the goal is to stay the same for a while. Until when? Simply, until your youngster looks right, looks healthier. Do not forget that you will be encouraging exercise and serving good, nutritious meals. It should not be long before all the family starts to glow with health!

What if your child falls into the second category and is moderately overweight? You have several options open to you. The first is simply to set a goal of staying the same weight (as for the first category). Alternatively, you might explain that the main goal is not to add weight (and this will be rewarded) *but* there will be an additional reward if the scales register a drop. The aim should be to lose 1 lb (half a kilo) or so every now and again; every second week would be enough. Watch out that your youngster does not get *too* enthusiastic about losing weight. Point out that a loss of just 1 lb (half a kilo) per month totals 12 lb (6 kg) in a year. A loss of 2 lb (1 kg) per month totals 24 lb (12 kg) in a year. This is going to cure all but the greatest overweight problems! Concentrate on positive aspects of good nutrition, not just restricting food.

It could be that your child falls into the third category, and is very plump. There is still no need to be super strict. Have a weekly weigh-in and give a weekly reward for the goal of not gaining weight. However, if there is a loss of weight give an extra reward. Do not expect your youngster to lose weight every week. A total of 2–3 lb (1–1½ kg) a *month* is quite adequate for most purposes. This amounts to a huge 24–36 lb (12–18 kg) lost over a year!

It could be that you start with a different problem. Your child is underweight. The most important thing is to concentrate on good, nutritionally balanced meals. As we explained in Chapter 4, good food need not necessarily be expensive and can be adapted to your child's taste preferences.

A simple goal for a very thin child is not to *lose* weight. Have a weekly weigh-in. Give a reward each week if your child is the same as the previous week. Give an additional reward if there is a slight increase in weight. It is very difficult to specify the rate at which weight should be increasing. It depends on the child's rate of growth and other factors. So, your best bet is to concentrate on giving a healthy diet rather than on specifying an exact weight which should be added each month.

Goals for diet change

In the last chapter we explained how important it is for your whole family to eat a high-fibre diet which is low in sugar, fat, and salt. You will already have planned a better eating regime by gradually introducing small changes to the everyday diet. All these details will be in your diary. There is no need, therefore, to draw up an additional, very detailed plan for your child. Pick on some simple goals which are particularly applicable and useful for your youngster. For example it might be that your child is especially fond of:

> tomato ketchup on meals
> eating sweets on the way home from school
> double portions of pudding at school lunch
> grandma's home-made cake

You will already have plans for dealing with some of these problems. The necessary steps will be written out in detail in your diary. However, some eating habits (like the large pudding eaten at school lunch) might be a particular problem that your child finds especially hard to tackle. It could be that some of the food preferences shown by your youngster are not shared by others in the family. These, then, might go into the goal list. Make these goals specific. Each child's list will look different from any other's. For the child who has the eating problems we have listed above, what might the goals look like?

First, let us take the problem with tomato ketchup. How much does the child have a week? A whole bottle? A half bottle? The first goal will be to halve the amount usually taken. If a bottle usually lasts a week, it now has to last two weeks. An alternative goal would be to have tomato ketchup every alternate meal (rather than each meal!). When this goal is attained, and a month or two has gone by, it will be time to add another goal to reduce the ketchup level further. However, do not forget that each child is allowed some treats. Tomato ketchup is fine as an *occasional* addition to a meal. It is important to work out the most appropriate goal (or series of goals) with your child. Friendly negotiation should be the order of the day.

Some problems will be trickier to tackle than others. Particularly difficult are occasions when food is eaten away from home. In the example we have given, school lunches, eating cake at grandma's, and having sweets on the way home from school are all hard to tackle. Does your child have problems like these? It is always difficult to ask your youngster to do something like refuse a favourite food, if you are not there to give moral support. It is always tempting to cheat a little—after all, who will find out? To help your child gain self-control in difficult situations, there are a number of steps to take. Make the goals easy. To begin with in our example, goals might be:

1. To halve the quantity of sweets eaten on the way home from school (or, have these sweets on alternate days rather than each day)

2. To have single portions of pudding at school lunch (eaten slowly!)

3. To have only a small piece of grandma's cake.

Discuss with your child goals which seem reasonable. In addition to this, it is essential that you explain the reasons why you want these changes to occur. Some simple education is necessary. Make this fun by making it into a game. If you have children without weight problems, they can join in as well. For example, take pictures of foods (or use real food) and use the traffic light colours to describe which is a good, 'go' food (green), an undesirable 'stop' food (red), and which foods are somewhere in between (orange or amber). Say *why* certain foods or drinks are good, and *why* some are bad. The amount of information you give and the way in which you present it will obviously depend on your child's age. You will need to give rewards for refusing the 'red' foods. But make it clear that good eating is not just about refusing food, it is also about having better alternatives. Examples are jacket potatoes instead of chips, fruit instead of sweets, wholemeal bread instead of white bread, and so on. Perhaps you can involve the younger members of your family with planning or even with cooking healthier meals of their choice.

As time goes on and goals are attained, it might be appropriate to set some new goals which will mean an even healthier diet. Just remember to always go at an easy pace. The aim is for your child to *achieve* goals and not get frustrated or angry.

Goals for eating-related behaviour change

You do not want to bombard your youngster with huge lists of goals. He or she will already have a goal for weight and some diet goals. Keep the goals for behaviour change few and simple. Perhaps you can negotiate just two or three of these. Gear them to the particular needs of your youngster. For example, if it seems that eating too quickly is a major problem, make a goal to eat slowly, savouring different tastes. Obviously you are not going to sit there with a stop watch! Set a good example yourself. You know how long it usually takes for the family to finish a meal. Make it longer. Stop to talk between mouthfuls.

Perhaps one goal could be about leaving food or taking smaller portions. The ideal will be for your child to think about when he or she feels satisfied—the goal is to stop eating as soon as that point is reached.

There again, a goal might be to stop eating food in any other room besides the dining room or kitchen (certainly, no food should be eaten in the bedroom!).

Another possible goal is to have sweet treats only once a day, or at a certain time of day. It is entirely up to you and your child to pick goals which are useful.

Rewards for attaining goals

Your child should now have a list of goals for weight, diet change, and eating-related behaviour change. The list should not be too long. Aim for a *maximum* of ten goals at any one time.

Now comes the fun when your child gets to see what rewards are available for attaining the goals. Rather than assume that you know what your youngster likes, sit down and ask for a full list. Include the obvious things like toys, trips out, parties, clothes, and holidays. Suggest to your youngster that the list can be as wide as he or she likes. Try to end up with at least ten or twenty possible rewards. Make sure the list is practical though, a trip to the moon is out of the question!

Obviously, if your child has ten goals to attain each week, there is a possibility of gaining ten rewards. Within a week or two the list will be worked through. Even assuming you are rich enough and willing enough to give all the rewards in so short a space of time, this would not be a good idea.

Each reward is likely to cost something in terms of money. You can therefore agree on a *small* amount of money as a reward for each goal, or you can buy or draw gold paper stars as a reward. Whichever you choose, the idea is to have a barter system. When a certain amount of money has been earned there will be enough to buy a toy or something else on the list. Alternatively, when a certain number of points have been acquired, or a certain number of stars put onto a

Weight control programme: diet

chart, they can be exchanged for a full reward from the list.

You need to decide how many points, or stars, or how much money you will give for each goal. (It is probably worth giving the goal for weight the most stars, points or money, whichever you are using.) Again, ask your child which goals look most difficult and give more for these.

Next you will need to work out an agreed exchange rate. With money this is easy. If a major reward is a toy which costs £5, then obviously the youngster will need to have earned this amount before being able to buy it. Try to work out a fair number of points or stars which need to be acquired in exchange for each reward. Now you are ready to start!

Should you ever use punishment?

The answer to this question is simply NO. If you start using punishment it will not help your child to attain any of the goals. In fact, it could well have the opposite effect. If your youngster senses strong disapproval from you, or feels hurt in any way, then he or she is likely to seek comfort from the food that is so often associated with warmth and love. A child gorging on sweets and chocolate is the last thing you want.

Another problem with punishment is that people seek to avoid it and this can obviously lead to telling lies. If your child has a triple helping of suet pudding at school lunch, it is the easiest thing to convieniently 'forget' about it. If telling the parents is going to result in punishment or verbal disapproval, this can easily be avoided by saying nothing. You obviously do not want this to happen. If your youngster is finding some particular goal hard to attain, you should not punish him or her for not trying hard enough. You should not nag or threaten. You should NEVER take back rewards (points, stars, money, or anything else) once they have been earned. Even if there is a certain amount of backsliding on progress, you never reclaim rewards. Not being able to attain *new* rewards because *new* goals have not been reached is bad enough, without you adding extra problems for your child. If your youngster is having difficulty in achieving a goal, you re-evaluate that goal. Is it too hard? Can you work towards this goal by setting up smaller, easier goals along the way? There will always be a way around a problem, but you will have to find that way by a process of trial and error. You will need a completely open, honest line of communication between you and your youngster in order to find a solution to diet and weight problems.

Monitoring progress

It helps a great deal if both you and your child can actually see progress from week to week. You should have a clear list of goals and rewards written out. Next to each goal will be the number of points or stars or money that can be earned by reaching this goal. You will need to note down on a separate sheet the 'exchange rate', how much in the way of points, stars, or money needs to be earned in exchange for each reward. Do not keep these lists on scrappy pieces of paper in the back of a cupboard, or in a notebook hidden away in a drawer. Write the information neatly on a couple of large sheets of paper. Perhaps you could pin them up on a wall or inside your child's wardrobe. Put them where they can be seen, or where your youngster can easily gain access to them. Tick off the goals as they are achieved. Make a graph to plot weekly weight. Use another chart to stick stars on or to keep account of total points or money gained. Go to some effort to make these graphs and charts colourful and attractive. If you make weight control into a game which is easy and fun you are sure to find success.

Summary

While it might have been tempting to think of trying to solve your child's weight problem with some kind of special diet, it should now be clear that this would not work. Rather, what you are looking for is the establishment of long-term healthy eating habits. However, in the short term, you will probably need to take some steps to get your child through the period of change from old habits to new. As far as diet is concerned, you will want to set up, with your child, appropriate goals of weight, diet change, and eating-related behaviour, together with suitable rewards. If you wish, you can use the tables we give to help set goals for weight, but more often it will be simpler just to look at your child's appearance. Set fairly easy goals to start with, moving on to the more difficult ones once the easy ones have been achieved. Do not try to force your child to change by using punishment—if you do, the change will be resented and will not last. It may even result in your child lying to you. Rather, you should aim to make the process of change into a game which both you and your child will enjoy.

9 Setting up a weight control programme with your child: exercise

We have already noted that a useful contribution to your child's weight control can be made by adjustments in activity level. Physical activity can have a number of effects on the body, of which at least two are very relevant to weight control, the reduction of fat and the increase of muscle. You will remember from Chapter 5 that this may mean that an overweight child does not necessarily become much lighter, but nevertheless develops a much better shape, as a large volume of fat is replaced with a smaller volume of more solid muscle. If, of course, your child is *underweight*, there will not be that much fat to lose, so here exercise can have the effect of increasing total weight by producing an increase in muscle.

Ideally, you want to start your child off on the right road by building up exercise habits that will last a lifetime. Regular exercise, for both children and adults, can result in a body that feels and looks good. On the other hand, to get to this stage from the starting point of a child with weight problems obviously requires a little planning and a few intermediate stages.

Step 1 Find out about your child's *present* activity level

As with planned change in eating habits, changes in exercise levels should start with a clear understanding of what is happening *now*. The diary in which you are keeping a check on eating can also be used to keep a record of your child's activity level. Exactly how easy this is to do in practice depends on a number of things. For instance, with older children you could explain to them in some detail that you want to help them organize more exercise, and that part of doing so is keeping a record of present activity. Indeed, some children may be able to keep a diary for themselves. For younger children you will almost certainly have to keep the record for them, and you may have to obtain some of your information by direct observation rather than asking.

The information you obtain may, because of these difficulties, only be a rough estimate of your child's activity. This does not matter, a rough guide may be all you need. The idea is to have at least some notion of where it is you are starting from, what are the strengths and weaknesses of your child's present activity programme!

This does not mean that you have to watch your child every minute of every day, keeping a continuous record of everything she or he does. Perhaps the easiest way is to divide each day into morning, afternoon, and evening, and note roughly how much time is spent in different levels of activity. You should then put each of the things your child does into one of four categories. These are:

1. Little or no activity, e.g. playing with toys on the living-room carpet, sitting reading or drawing, etc.

2. Light activity, e.g. wandering round the school playground talking to friends, strolling to the local shop with you

3. Moderate activity—this may be something which is fairly active and continuous, e.g. a long brisk walk with few stops, or something which is a mixture of high and low levels of activity, e.g. a game of netball with sprints up and down the court interspersed with standing around waiting for the ball

4. Intense activity—anything which keeps the child moving fairly vigorously for a continuous, sustained period, e.g. kicking a football about with one other friend (if there are only two children there is little chance to stand around doing nothing!), steady swimming at the pool, etc.

Your diary should then be able to note roughly how much time during each day is spent on each of the four kinds of activity. In order to complete your diary you may well have to ask your child about what has been done at school, etc. It is probably no bad thing to ask about your child's day; after all, if you have been away from your partner all day you will typically ask (and expect to be asked) about the sort of day they have had, yet rarely do parents ask their children the same question. On the other hand, you must be careful not to make it into an inquisition—ask in a polite and friendly way, explaining that you are *interested* in the child's day. Under no circumstances should what you have been told result in your child being scolded, teased, or in any way made to regret having told you. Remember that the whole point of all this is eventually to make your child happier as well as healthier!

Keeping the record need not involve you in too much effort as it does not have to be incredibly detailed—all you are after is an idea of the pattern, amount and intensity of your child's present activity level, so as to see what changes may be most appropriate. Thus, a typical week's diary may look something like this:

Monday	Morning; school, 15 minutes sitting around and strolling about in the playground (Level 2) Afternoon; reading during lunch break, rest of afternoon as morning (Level 2) Evening; homework, watching TV (Level 1)
Tuesday	Morning; school, as Monday (Level 2) Afternoon; as Monday (Level 2) Evening; as Monday (Level 1)
Wednesday	Morning; raining, indoors all morning (Level 1) Afternoon; reading during lunch break, 90 minutes hockey in games period (Level 3) Evening; stroll to friend's house after homework, sit chatting, stroll back (Level 2)
Thursday	Morning; as Monday (Level 2) Afternoon; as Tuesday (Level 2) Evening; as Monday (Level 1)
Friday	Morning; hour's PE class, quite active (Level 4) Afternoon; as Monday (Level 2) Evening; as Wednesday (Level 2)
Saturday	Morning; in bed (Level 1) Afternoon; strolling around shops with friends (Level 2) Evening; disco at youth club, 2–3 hours dancing (Level 4)
Sunday	Morning; in bed (Level 1) Afternoon; playing board game (Level 1) Evening; watching TV (Level 1).

A diary like this is by no means unusual, yet it reveals

that in a seven-day period there is only a tiny percentage of the time spent in activities beyond Level 2. Out of nearly 170 hours in the week, only about five are spent in brisk activity. Even allowing for the time spent sleeping, say eight hours a day, this still means that less than five per cent of the child's time is spent in reasonably active pursuits. Clearly, there is room for improvement. What you should aim for is to have at least 10–15 minutes of Level 3 or Level 4 activities most days. This might involve, say, jogging a mile or so round the block, or swimming 10 or 15 lengths of the swimming pool. If activity stays below Level 3, at Levels 1 or 2, then it takes much longer to get the same value from the exercise—perhaps an hour or more, for example, of walking at normal pace. Ten or 15 minutes a day of Level 3, however, is likely to be a lot more than your child is doing now.

Before setting your longer term goals you will need to keep a record like this over a couple of weeks. Only then is it time to try to organize some changes.

Step 2   Building up the activity

Having obtained a picture of your child's activity level, the next step is to try to bring about some subtle changes. What you want to do is to increase both the *quantity* and the *quality* of your child's activity. As far as weight control is concerned, the kind of activity you are going to be most interested in is that which involves fairly sustained, steady effort. A good level of exercise is one in which, for most of the time, your child is breathing more heavily than normal, but at the same time still able to hold a conversation. If breathing is normal, the chances are that the exercise is too mild to be of much benefit; if breathing is too heavy, the exercise is probably uncomfortable and will be impossible to sustain for a long enough period.

Weight control programme: exercise

Let us look at a diary such as the one given earlier. How might we try to bring about some changes in exercise level? Perhaps the first thing to notice is that there are a couple of occasions in which this child is being fairly active without being forced; for example, two or three hours at the disco suggests that disco dancing is fun, and the fact that the PE class was quite intense may imply that this kind of activity is also enjoyable (unless the PE teacher is remarkably good, the odds are that children who do not enjoy PE find a way to do as little as possible; the fact that this child did participate suggests that the activity is actually considered to be quite fun).

There is very little that you can do to alter the school timetable in order to get more PE classes. If your child is overweight, the chances are that, despite enjoying the class, he or she still is not one of the best at it. This means that there is probably not much chance of adding anything extra to what the child does at school (if the child is good at gymnastics, it might be possible to join in the school's out-of-hours gym club).

What this means, we're afraid, is that *you* have to think about taking an active part in building up activity. First, you need to talk to your child to find out exactly what it is about any particular activity that makes it enjoyable. Is it the rolling and tumbling of the PE class? The opportunity to be with friends at the disco? Similarly, what is it that makes, say, the hockey not so enjoyable? Difficulty in controlling the ball with a hockey stick? Not being able to sprint fast enough? Finding out about what makes the difference between enjoyable and unenjoyable exercise is a useful first step towards building up extra activities.

The second thing to do is to take the things which make activity more or less enjoyable and try to find things which fit similar patterns. For example, if the child likes the stretching exercises of a PE class, and the disco dancing of the youth club, would it be fun to have a go at aerobics? Going out to an aerobics class, even if you can find one suitable for children, is possibly too frightening for a child who is aware of not being in the best of shape. Yet there is nothing to stop you obtaining one of the many aerobics workout records, clearing a space in the living room and running through the routines at home. *But*, and here we have an important message for all exercise planning for children, doing so will probably mean that you have to do it as well. Yes, it is probably necessary for you to share the exercises with your child if you are going to maintain motivation. Simply telling your child to get on with it alone will just result in feelings of being put on, singled out, and generally persecuted; allowing your child to share in your exercise will quite possibly make them feel special, and conceivably cause them considerable amusement!

Indeed, you may find that besides keeping a record of your child's activity in the diary, it is not a bad idea to keep a record of your own and the rest of the family's. It can be quite sobering to see just how little healthy exercise each of us actually takes. If you want your child to grow up with a fit, healthy body and the habits to keep it that way, you will probably have to set a good example yourself!

As a general rule you will probably find that taking part in some physical activity yourself is a good way to encourage your child to do so. There is a lot to be said, therefore, for considering the kinds of activities which both you *and* your child can enjoy. The kind of activity best suited to weight control is, as we have seen, the kind which involves mild, sustained effort. Some possibilities for such exercises, which both you and your child can do together, include:

1. The kind of *aerobics* or similar workout we have already mentioned. Look at the records and cassettes available in the shops, pick something which appears suitable, and

set aside time to go through the routine regularly with your youngster. If you can afford it, suitable new clothes (track suit, leotard) can make a nice present to encourage both of you to start.

2. *Cycling.* Although buying a bicycle can be a fairly substantial outlay, there is every possibility that it can also save you money in petrol, bus fares, etc. Your child may indeed already have a bicycle, and you may have one yourself. Bring them out of whatever storeroom, garage or garden shed they have been consigned to, clean them up and make sure that they are safe, then organize regular use. Such use can range from trips to the local shops to expeditions to places of interest around the area you live. Study local maps and guidebooks to find suitable areas for picnics or exploration—country parks, the local zoo, or other places of interest. Consider the possibility of taking bicycles part of the way on a train, then exploring your destination on two wheels; many trains in Britain will carry your accompanied bicycles free of charge.

3. *Swimming.* The overwhelming majority of people live within reasonable distance of a swimming pool. A regular Saturday or Sunday morning swim for the whole family can give extremely good exercise that is fun to do. When you go to the swimming pool, make sure that you are actually *with* your child—do not go speeding up and down the pool leaving your youngster idly splashing about in one corner. Swim together and devise games to play together. Take turns in swimming to each other from opposite ends of the pool. Practise swimming underwater. See how many widths, or lengths of the pool you can do each week. Make sure that once you have made it to the pool, you are actually swimming for most of the time that you are there.

4. Other family *sports and games*. There are a number of ways in which you can get some or all of the family involved in physical activity. You and your child can play informal football in the back garden; with just the two of you, there should be little chance to take things easy! Several of you can find a little spare ground and take a 'frisbee' or similar item to throw around—it is surprising how much energy is involved in something like this. Long brisk walks through country areas, or even running through the park while walking the dog, can all be ways of increasing your child's total activity level. Remember that there is benefit to be obtained from any exercise that is regular, sustained, and mild.

There is no reason, of course, to restrict yourself to only one of these activities. Indeed, several may be combined to good effect. Cycling to the swimming pool, for example, enables you to fit two forms of exercise together quite naturally. Even where activities do not go particularly well together on any one occasion, it is still possible to do one activity today and a different one tomorrow. Of course, what we have given you are only a few of many possibilities open to you. You may already have particular interests of your own in which you can involve your child. You may go ballroom dancing, or jogging, or play football or netball with a local team. If so, it is worth thinking about letting your youngster share with your practice. Again, clear your living room and put on a record to practise dancing. Buy your child a good pair of training shoes and take them jogging with you. Kick or throw a ball around with your child in the back garden. By and large, children love to share in adult activities; it gives them a feeling of being grown up, and part of the adult world.

Setting goals

Ideally, your child should reach the stage of doing some Level 3 or Level 4 activity six or seven days a week. As with the changes in eating habits, however, this has to be done gradually (if you are not used to exercise either, doing it gradually may be important to you as well!). Just introducing a single weekly session of aerobics, cycling, or swimming may be all that is appropriate for the first couple of months. Remember that your child's eating habits are going to take at least a year to change. To spend the same time building up your child's activity level, even if it starts at only one day of exercise a week, still gives you plenty of time to build up to seven days a week within the year. If you manage Level 3 or Level 4 exercise one day a week for the first eight weeks, then two days a week for the next eight and so on, by the end of the year there will be regular exercise on six days a week. For most children this is certainly as much as they will need; it may even be better than seven days a week for the majority of children, a day of rest doing everyone some good!

Because you have so much time to work with, it is not actually necessary to start out each time with a Level 3 or 4 activity. For example, let us suppose that your child is currently at the stage of only having one day a week during which any Level 3 activity occurs. Your first goal need not be to have two days a week on which there is such activity. After all, there may be some days on which there is hardly any exercise (like the Sunday in the diary example we gave earlier). A first step to building up to two days of useful exercise a week might be to try to introduce some Level 2 exercise to such a day. This might involve, for example, a trip to the swimming pool, where most of the time is spent splashing around and playing games, or a cycle ride at a leisurely pace with frequent pauses. Often it may be worth spending a few weeks at this easier level before gradually increasing the intensity to Level 3 or even 4.

Another point to consider when allowing time for goals is that, in general, progress might be slowest when your child is first starting. It is usually much easier to progress from five days exercise a week to six than it was from one to two, even though both are adding one more day. This is partly because going from one day to two days is a 100 per cent increase in the exercise, whereas going from five to six days is only a 20 per cent increase; and it is partly because by the time a child is doing five days exercise a week he or she will be much more fit, and therefore more able to increase the exercise. What this means is that in planning your year's goals you may not need to allow as much time for progress later in the year as you do for the early goals. Thus, during the year you may allow something like the following amounts of time for your goal planning:

1. Level 3 or 4 activity for at least one day per week; 10 weeks

2. Level 3 or 4 activity for at least two days per week; 10 more weeks (20 weeks total)

3. Level 3 or 4 activity for at least three days per week; 9 more weeks (29 weeks total)

4. Level 3 or 4 activity for at least four days per week; 8 more weeks (37 weeks total)

5. Level 3 or 4 activity for at least five days per week; 7 more weeks (44 weeks total)

6. Level 3 or 4 activity for at least six days per week; 7 more weeks (51 weeks total).

Thus, in just under a year it is possible to move from nothing but minimal exercise throughout the week to moderate activity six days per week, with never less than seven weeks to get used to the new level. Remember, too, that to a child these seven to ten week blocks will seem much longer than they do to you—two months or more can seem like an eternity to a child.

During these blocks you may wish to be making smaller adjustments on the days other than the one programmed for the increase. For example, in moving from goal 3 to goal 4 in the list above you may be trying to build up the habit of jogging half a mile or so on Friday evenings. While you are doing this you may also be introducing some Level 2 activities to another day of the week, for example an evening stroll round the block with the dog on Tuesdays. By introducing these additional elements of mild exercise, the nine weeks spent consolidating the Friday night run can also be spent building up a little activity on another night, in preparation for the next stage.

As with the setting of goals for eating habits, you must, of course, be flexible. Some goals may be particularly difficult to attain; you may find, for example, that there are just so many attractive alternatives that your child fails to take moderate exercise nearly every day. Do not force things. It may be that as your child comes to enjoy the exercise more, the activity may automatically become preferable to the competing alternatives. Or you may find that it is easier to increase the amount of time spent doing moderate or intense exercise on four days a week than it is to add such exercise to an extra day. You may of course find that your child is already doing a number of fairly active things. If so, that means you can spend more time over the year on each stage of the build up. If your child is already active during three days per week, you have only three stages to go through to build this up to six days a week—so you can spend four months or so on each stage! Do keep monitoring your child's activity throughout the year, as you may find that besides the extra activity which you are encouraging there is a greater tendency to do active things at other times as well, which, of course, is very encouraging. In many ways building up the exercise is very similar to programming the diet change. As with diet change, you should look at your child to assess progress. Is your youngster a better shape? Is there less fat, or more muscle, or both? Use your child's appearance to guide you in setting and adjusting targets. And above all, make it fun!

Suppose you cannot join in?

Of course, it may be the case that you are unable, for one reason or another, to join in with much of your child's activity. You may, for example, have some handicap or medical problem which restricts your own capacity for physical activity; if you think this is the case, work out carefully what it is that you would like to join in with (a cycle ride, a trip to the swimming baths, or whatever) and check with your doctor to see if it is a good idea. For the level of exercise that we are considering (not less than ten minutes or so at a level producing slightly heavier breathing than normal) it is quite possible that your doctor will agree to your joining in. If not, try to obtain advice from your doctor as to what exercise you may be able to do.

Much more commonly, you may be unable to join in with your child's exercise simply because it is inappropriate or impractical. If, for example, your child wanted to join the local judo class, you would probably find that membership of the class was restricted to under-16s; even if you could join the class you might find yourself a little out of place amongst all these youngsters! Similar problems apply of course to many sporting and similar activities, such as ballet classes.

Where this occurs there are still certain things you can do

to help. First, you can check to see whether there is a similar class, perhaps in a different room or at a different time, that you *could* join in with. Thus, even though you are not actually with your child during the exercise, there is still an element of sharing involved, enabling the two of you to talk about your experiences, successes and difficulties. Secondly, and especially relevant if nothing would induce you to take up your child's chosen activity, you can be as supportive as humanly possible to their efforts. Discuss with your child what goes on in the session—how the exercise is going, what is difficult and what is easy, who else is there, and so on. If appropriate (i.e. if your child wants you to) provide transport to and from wherever the activity takes place. Listen to your child's needs for special clothing or other items, and provide the essentials as quickly as possible—perhaps provide other items as part of a reward system like the one you use for food. If there are competitions, examinations, or similar events involved, be proud of your child's successes and sympathetic when things go wrong. Even if you cannot take part physically, share as much as possible of the activity with your youngster.

Step 3 Making it regular

Once you have introduced some extra activities your next task is to try to make them a normal part of your child's routine. In general, the best way to do this is to make sure that the activity is enjoyable. When your child first starts to go on cycle rides, swimming, or whatever, part of the enjoyment will probably be coming from the fact that it is shared with you. However, in the longer term, you want the activities to be so enjoyable in themselves that the child will do them whether or not you are joining in. This will take time. By and large, the child with weight problems will not, on first starting, get sufficient enjoyment from the activity itself to provide adequate motivation to continue. That is why you will be joining in. But later, as skill and fitness improve, the exercise itself should become enjoyable in its own right. Moreover, the confidence your child obtains may well result in a keenness to try out other forms of exercise. Some of these may be ones you would never dream of doing—attending a ballet class, a judo class, or learning the triple jump, perhaps! Do not worry—if you reach the stage where your child is asking to take part in these activities, the odds are you will have established the exercise habit.

The most important aspect, both of using exercise to control weight and of establishing the exercise habit, is to make sure that the activity or activities you choose are done regularly. In the diary example we saw earlier, the child in question actually did Level 3 or Level 4 exercise on only three days out of the seven in the week. On two of these the exercise was part of the school day; the odds are that during school holidays nothing would be organized to replace these and exercise would be down to one day per week. Remember that for a third of the year, or so, children are not actually in school, so that relying on school games and exercise periods leaves your child with little in the way of exercise for much of the year.

To try to build up regularity in exercise, therefore, you will need to do several things. First, and most importantly, you will want to make it enjoyable for your child. If your youngster is not enjoying the exercises, good habits will never get established. Secondly, as far as it is compatible with your child's enjoyment, you will want something that parents can enjoy as well! Providing a compromise between these two aims will require a degree of sensitivity on your part, together with some firmness. It is no use either forcing your child to do something you will enjoy but he or she will not, or forcing yourself to do something you do not enjoy just for your child's sake. Talk to your child, point out (if appropriate) that *both* of you could do with more exercise, and negotiate what form this will take. Try, as far as possible, not to interfere with your child's normal enjoyments—you cannot expect a jog round the block to be enjoyable if it means that your child is missing a favourite TV programme.

You may also want to think of other ways to increase your child's motivation. Buying presents to do with exercise, as long as it is not to the total exclusion of other presents, can increase motivation. A smart new tracksuit can lead to an urge to go out running; a pair of table tennis bats and a ping-pong ball can result in the dining room table being cleared for action. A new swimsuit or goggles will soon have to be tried out. Things like this can serve both to start your child off and to help maintain enthusiasm. You do not necessarily need to use these to reward particular achievements (e.g. swimming the whole length of the pool); it may be enough just to give a reward for sticking at the chosen activities.

Coping with problems

Needless to say, it is unlikely that everything will proceed entirely without a hitch. Various problems can arise in trying to bring about changes in exercise level, which will need to be dealt with if weight control is to be efficient. Such problems may include the following:

● Your child does not enjoy the exercise. If your child's motivation is low, then it is a fair bet that for some reason things are not enjoyable. If this is the case, the first thing to do is to try to be clear about whether it is the exercise itself that is not enjoyable or whether it is something about how it is organized. If it is the exercise itself that is not enjoyable, there are two things you can do. First, you can try to find out from your child if there is some other form of exercise that is more fun, and switch to that if possible. Try to find out if there are exercise-related toys or presents that your child would like—a table tennis set, a punchball, a skipping rope, or something similar. If you can buy such a present, the odds are that the novelty of the present will provide motivation for a while at least.

Secondly you can try to boost your child's motivation by means of other presents, encouragement, praise, and so on. Very often exercise is not terribly enjoyable at first because it is difficult and unfamiliar, but as your child becomes more accustomed to it, it may well become enjoyable in its own right. In situations like this, your job is to try to provide enough encouragement to keep going through the early stages.

Do try to find out, though, whether it is the exercise itself that your child does not enjoy, or whether it is something to do with how the exercise is programmed or organized. If your child is having to miss a favourite TV programme, try to reschedule the exercise, or to videorecord the programme for later viewing. If your child is being teased about the exercise, try to have a word with those who are teasing. If your child already has lots of other enjoyable activities which the exercise is taking time from, try to look for activities of fairly short duration—for example, a ten-minute jog rather than an hour's walk. It really is worth putting a little imagination and effort to make your child's (and your own) exercise enjoyable.

• Your child does not maintain any single activity for more than a few weeks. Again, there are two aspects to this sort of problem. It may be that your child is simply having difficulty finding an activity that is enjoyable, despite persevering for the first few weeks to see if it gets better. If so, such perseverance deserves some credit. Alternatively, it might be that the novelty of the activity keeps it interesting for a few weeks or so, after which your child becomes bored. Such boredom is often associated with parental boredom, and the first thing to ask in such circumstances is whether or not you are still taking the same interest and giving the same encouragement as when you started. If not, then this is a useful reminder to keep up your own involvement. On the other hand, it may simply be that the activity really is not as interesting—but this need not be a problem. There are many ways of taking exercise, and there is a lot to be said for varying them anyway. So if your child really cannot be shifted from this tendency to change activities every few weeks, do not make a fuss, simply encourage the new activity and keep thinking of others! If you do eventually run out of ideas, it is often going to be the case that sufficient time will have elapsed since your first idea to be able to go back to the start again with renewed motivation.

- You cannot keep up with your child! The opposite problem to the previous ones, it may be that your child takes to exercise so well that joining in exhausts you. Although it might be something of an eye-opener for you to realize that you cannot keep up with your youngster, it is not too much to worry about from your child's point of view. For your child to do better than you at some form of exercise will require at least a reasonable amount of motivation, so such success will usually mean that he or she will manage fairly well with your absence from some activities. You may, however, want to use this as a sign that you, too, could do with building up your activity level.

- Your child is overweight but too young for exercise. Actually this is very rarely the case. Even a crawling baby is using up a fair number of calories in getting about the place, and the typical toddler, given free rein to run around, will be doing easily as much exercise as is needed. If you simply spend time playing physical games with your baby—holding your child's hands to help with standing upright for example, or rolling a ball around the living room carpet—the odds are that you will be doing quite enough in the way of exercise; any further effort and changing weight for such a young child should be done through looking carefully at the diet.

- Your other children take exception to the over- or underweight child's special treatment. And well they might! The point is, of course, that any special treatment given to one child should certainly be paralleled by similar treatment to other children. If you are buying one child a present as a reward for success at some physical activity, make sure that other children have their own opportunities to obtain similar rewards for things they need to be encouraged to do. Often, of course, it will be possible to apply exactly the same sort of programme to all the children in the family, without singling out just one of the children because of a weight problem.

- You do not have time, or you cannot afford it. We will accept that you may not be able to afford the time or money if your child asks for a month's skiing holiday in Switzerland. More often than not, however, complaints about not being able to afford it are less true when you look more closely. Some of the things you might feel it necessary to buy for your child's activity do not necessarily involve any long-term expense; if you buy your child a tracksuit, for example, then whenever that is being worn, other clothes are not, so you are saving money on the latter (indeed, something like a tracksuit is so baggy that you will probably find that it is not grown out of as quickly as more closely fitting clothes, and that it ends up saving money). Of course, some items will involve you in special expense, but this need not be excessive and, after all, no one ever said that bringing up a child was cheap!

As far as time is concerned, again you must expect to have to give up some of *your* pleasures to involve yourself in your child's activity; but after all, isn't your child also one of your pleasures? Anyway, as we mentioned earlier, the time involved need not be substantial—to do a mile-long jog around the block will take less than quarter of an hour out of your day.

Some DOs and DON'Ts for your child's exercise

DO try to reach a stage where exercise is taken *regularly*—a little moderate exercise on a regular basis is much better than intermittent intense exercise.

DO join in with your child's exercise wherever and whenever it is possible and appropriate.

DO remember that your child should enjoy the exercise.

DO build up the exercise habit gradually.

DO give lots of encouragement.

DO make sure that exercise is adequately supervised, either by a teacher, a parent, or some other responsible adult. This applies to sports and games, and to activities where there are other risks, such as jogging on a main road or in the dark.

DON'T just tell children what to do and then leave them to get on with it.

DON'T ask children to do any exercise that is dangerous.

DON'T put children into activities that are highly demanding—by all means let your child jog for a few miles, but do not let young children do things like training for the marathon—excessive training can put growing bodies at risk. Similarly, beware of other activities which put a heavy strain on the body, like weightlifting.

DON'T compete with your child or use the exercise to show off how much better you are than them—this will only discourage or bore your child.

DON'T encourage exercise if your child is ill, excessively tired, or injured.

Summary

In this chapter we have outlined how you should go about including exercise in your child's weight control programme. This should involve stages of assessing your child's starting point, gradually building up the level of activity, and making that activity a regular habit, so that your child gets at least ten minutes or so of continuous moderate exercise most days. Take your time doing this, allowing a year or more to build up the new habits. To boost motivation, you will probably need to become quite involved yourself, either by taking an interest and providing lots of encouragement, or even by joining in yourself. Done carefully, a programme which builds up exercise can not only set the pattern for a lifetime's healthy behaviour, it can also give a great deal of pleasure.

10 What happens once your child's weight is under control?

Over a period of a year or so your child's weight should come slowly, but surely, under control. We have given you lots of advice on diet change and the importance of exercise. It is also essential for you to develop the right attitudes—doing things which promote health should be more fun than carrying on with the old unhealthy habits.

We have given you ideas on many areas for possible change. For example, the whole family's diet can be improved. You can develop special programmes for diet and exercise with your child. The family can be encouraged to be more energetic and to enjoy active leisure activities together. You can try some simple nutrition education with your child (or children). Mealtimes can be a real social event; take time to talk and enjoy the food. Meals should be more pleasurable than ever before—you will be eating more slowly, savouring new tastes and textures. Your child should be more aware of the real feeling of hunger, the joy of eating good food to satisfy that hunger, then *stopping* when full. Food should no longer be nibbled at mindlessly, without real enjoyment.

As you can see, there are many options open to you. Do not worry if you feel you cannot manage all the areas of change at once, or if some of the things you try are not completely successful. The essential thing is to keep at it, slowly making improvements. Every family and every child is different. It could be that in your particular case one area is easier to manage than others. For example, you might find that changing food is difficult, but getting your child to exercise more is far easier. Or the opposite may be true, exercise change being hard but food change being easy. There again, you might find that making small improvements to the whole family's food is easy, but trying a special programme of diet change just for your child is hard. Someone else might cope perfectly well with a special programme for the child but find that changing the whole family's eating habits is too difficult.

Get going on the changes you *can* make. Keep a record of progress in your diary as we have shown you earlier. You may well find that as time goes on your child's weight comes under control by just concentrating on one or two areas of change. For some children, simply encouraging exercise can have a big effect. For others, just cutting down on sugar (or fat), while increasing fibre can make a big difference. There again, a major thing in some families might be a change in attitude—if you never press a child to take food that is not really wanted, if you stop insisting that all food be eaten up at mealtimes, if you do not offer sweets—this can have an enormous effect on weight. Do what you can and do not give up. Do the easiest things first, and take your time introducing the harder changes.

Although we have given you lots of ways in which to influence your child's physical shape and encourage good weight control, we have *not* given you an option for fast weight change. Take a year or more to help your child. You might like to compare rate of weight change with rate of increase in height. Do you see your child growing taller? You certainly do not see a change from one day to the next. If you were to measure *very* carefully, you might just see a change from one month to the next, but the increase in height would still be very tiny. And yet, from one year to the

Once your child's weight is under control

next there seems an enormous difference. The warm weather comes, your child tries on last summer's clothes, and they are nowhere near long enough. Change in weight and body shape is a bit like this. It is not something that happens overnight. There does not seem to be much difference from one week to the next. Yet, over a year (spent introducing healthy eating habits and encouraging exercise) you will see an enormous improvement!

When should you stop keeping your diary of food changes? When can you throw the charts and graphs away? Basically, you should keep on with these things until they stop being useful, when you have achieved all your goals. They are there to help, to keep you on the right track, to show you and your child that progress is, indeed, being made. Because weight control is a slowly acquired skill, you will not see big, overnight changes. If your child's physical shape takes a year to change, you will need a written reminder that progress is actually being made on a daily basis. All the tiny improvements from day to day will eventually add up to a noticeable difference. By the time this happens, healthy eating should be second nature. When you have been preparing better meals (shopping for fibre-rich foods low in fat, sugar, and salt) for months, you will not need a diary to help you keep track. If exercise and games become fun for your child, you will not need to give extra rewards for pursuing them. Your child will look healthier, be lean and fit. Clothes will fit better. Compliments will start rolling in. These natural rewards should replace any other reward system you might have been using to ease progress along.

Never forget, too, that by giving your child good food and encouraging exercise you are setting the foundation of a healthy adult life. Develop and maintain appropriate eating and exercise habits and you are much less likely to suffer the major killer diseases, like cancer and cardiovascular disorders (strokes and heart attacks).

As a final note, have you considered what a major part eating plays in our lives? If a person lives to be 80 years old, and has, on average, three meals a day, this adds up to almost 88 000 meals in a lifetime. This *should* mean 88 000 opportunities for enjoyment! All too often, however, meals are not pleasurable. Perhaps the food itself is disliked. Maybe eating is a hurried affair, with little thought given to it. Perhaps meals are not sufficiently nutritious to be satisfying, or are eaten out of habit rather than as a response to true hunger. One of the worst things is when people cannot eat without feeling guilty. How many people in our society never really enjoy food because they feel too fat and therefore regard food as 'the enemy'? What a sad state of affairs this is. If we would only take *time* to establish good eating habits, we would find that weight control can become a natural way of life. If we want our children to be in good physical shape, it is important to teach them how to appreciate good, nutritious food, and to encourage exercise as part of everyday life.

Summary

Following the guidelines we have given should result in your child's weight coming gradually under control over a period of a year or so. Of the various ideas we have given you, some may turn out to be fairly easy to apply, some more difficult. For the beginning, at least, concentrate on the easy changes; these may be enough on their own to achieve the results you want. Eventually, the natural rewards of a healthier, fitter body will make it possible for you to discontinue your own, artificial rewards. If things go according to plan, you will have given your youngster a pattern of healthy diet and exercise that will last a lifetime.

Further reading

Cannon, G. (1986) *Fat to fit*. Pan books, London.
Graham, L. (1986) *A good start; healthy eating in the first five years* Penguin, Harmondsworth.
Walker, C. and Cannon, G. (1984). *The food scandal*. Century Publishing, London.

Index

activity, in children, *see* exercise
age, and weight 6–8, 105–9
  babies 4
alcohol 14
arthritis 4
athletes 59–60

babies 4, 74–9
  breast feeding 74–6
  weaning 76–9
  see also infants
baby foods 76–7
beer 15
blood pressure, and salt 52
breakfast 55, 87, 90
breast feeding 48, 74–8
  problems in 75–6

calcium 5, 42–4
  importance of 44
  see also milk
calories 37–40, 60
  counting 39
  definition of 37–8
  and exercise 63–4
  and weight 60–2
cancer 4, 139
carbohydrates, *see* sugar
cereals 51, 90
cheese 9, 40, 44, 57
  and babies 78

children
  activity levels in 117–29
  confidence in 68–9, 130
  difficult 20–1, 80–1
  means of exercise 64–5
  parents and, *see* parents
  putting into categories 31–2
  typical day's food 84
  weight control programmes 102–15
chips 48, 80
chocolate, *see* sweets
Christmas 25
confidence, in children 68–9, 130
convenience foods 91–2
crisps, frequency of eating 95
cycling 123

dairy products 47–8, 93–4
desserts, *see* sweets
diabetes 4
diet 4–6, 8–9, 13, 19–20
  changing 109–11
  changing, plans for 82–101
  controlling 17
  and exercise 69
  healthy 36–58
  wrong ways to 10, 14–15
  see also food
drinks, non-milk 24–5, 54, 87–9; *see also* milk

Index

eating, *see* food
eggs 45, 57
elderly 4

exercise 9, 12, 17, 30–2, 59–70
   do's and don'ts 135–6
   importance of 59–60, 67–9
   means of, by children 64–5
   motivation 131–3
   over- 133
   painful 66–7
   regularity of 67, 117–20, 123–31
   suitable 65–7
   types of 62–4
   and weight 60–2
   and weight programmes 116–36
   see also sport

family, *see* parents
fat 53
   in cooking 47, 94–5
   and NACNE 46–9
   reducing intake of 92–3
fat children, *see* obesity
festive food 25
fibre 6, 16–17, 20–1, 34, 51–2
   foods containing 98–9
   increasing dietary 98–9
fish 45, 56–7
food
   attitudes to 16–17, 27–9
   for babies 76–9
   calories 37–40
   convenience 91–2
   difficult children 20–1
   dislikes 19
   fibre in 51–2, 98–9
   habits and 27–9
   in illness 32–3
   for infants 79–80
   iron in 43
   and NACNE 46–52
   obsession with 34
   processed, and salt 52
   quality 36
   quantity 37
   refusal 27–9, 31
   right attitudes to, 17
   salty 96
   at school 23–4
   as a social phenomenon 14–15
   on traditional occasions 25
   typical, for a child 34
   in weaning 76–9
   wrong attitudes to 16
   see also diet
fruit 45, 53
   babies and 78

gall-bladder disease 4

habits, eating 27–9
headache 3
heart disease 4, 139
height, and weight 7–8, 105–9
herbs 54, 56
hobbies, *see* sport
holidays 25
honey, *see* sweets
hunger 14–15, 17–19, 26, 28–9, 137

ice cream, *see* sweets
illness 32–3
infants, feeding 79–80; *see also* babies
iron 42–4
   foods containing 43

Index

jams, *see* sweets

lemonade 11, 20, 24, 38; *see also* drinks, non-milk
lifestyle, setting of 9, 12–13, 26, 35

margarine 43, 45–6, 49, 53
meals
  breakfast 55
  eggs 57
  fish 56–7
  ideas for 54–7
  main 55–6
  recording of 83–6, 139
  school 23–4, 110
  social phenomenon of 14–15, 140
  timing of 32
  traditional occasions 25
meat 46–7, 94
milk 5, 40, 44, 47
  babies 75–6
  breast 48, 74–6
  skimmed 48, 57
  slimmers' 14
mimicking 12
minerals 42–4
money 112–13

NACNE, recommendations 46–52, 58
  fats 46–9
  fibre 51–2
  salt 52
  sugar 49–50
National Advisory Committee on Nutrition Education (UK), *see* NACNE, recommendations

New Year 10
nutritional needs, of children 37–46; *see also* diet
nuts 95

obesity, in children
  calories and 37–40
  deciding on 6–8, 105–9
  exercise and 64–5
  families 15–20
  observing 3–6
  weight, and height 105–9
  see also weight problems, in children
obsession, with food 34
  calorie counting 39–40

packed lunches 23–4
pain, and exercise 66–7
parents, and obesity
  assessing fatness 6–8
  attitudes to diet 4–6, 18–19, 27–35
  child mimicing of 12
  child relations 69
  children, and weight control 102–115
  deciding what to do 8–11
  participating in helping 11–13, 122–4
  sport 122–4, 128–9
parties 22–3, 25
potassium chloride 52, 97
presents, of confectionery 24–5
proteins 40–2, 53
punishment 113–14
'puppy fat' 4

refusing food 27–9
resolutions, making 10–11

Index

rewards, *see* treats
Ruthmol, *see* potassium chloride
salt 6, 34, 52, 54
  and babies 77
  in cooking 97
  foods high in 96
  recommended intake, NACNE 52
  reducing intake of 95–8
  table 97–8
school meals 23–4, 110
snacks, non-sweet 50; *see also* sweets
socializing
  eating and 14–15, 25–6, 140
  parties 22–3
sodium chloride, *see* salt; *see also* potassium chloride
sport 3–4, 9, 12–13, 30–1, 62–4, 121–4
  avoidance 3–5, 65–6, 120–2, 131–3
  parent non-participation 128–9
  parent participation in, 122–4, 128–9
  personalities 59–60
  see also exercise
stamina 63
strength 63, 68
sugar 6, 9, 16, 24, 49–50, 53
  NACNE recommendations 49–50
  reducing intake 86–8
  see also sweets
suppleness 63–8
sweets 6, 9, 11–14, 17, 19–20, 22–3, 25, 29, 38–9, 48–50, 86–92, 95
  for babies 79
  as presents 24–5
  see also sugar
swimming 12, 32, 123

teeth 4, 24, 42
thin children 19, 28, 108–9
treats 6, 11, 22–3, 110, 112–14

vitamins 6, 34, 43–6
  tablets 44

weaning 76–9
weight control programmes 102–15
  age and 73–82
  and exercise 116–36
  starting 82–6
weight problems, in children 5, 14–15
  after solving them 137–40
  age and 6–8
  attitudes 3–4, 18
  comparisons 6–7
  delay in tackling 8–11
  discerning 3–13
  easy ways of tackling 10–11
  and exercise, *see* exercise
  health and 4
  inherited 15–17
  parents and, *see* parents, and obesity
  programmes 104–9
  psychology of 3–4, 14–15
  resolutions 10–11
  teasing 3
  thin children 19, 28, 108–9

yoghurt 40, 44, 47, 54; babies and 78–9

MORE OXFORD PAPERBACKS

Details of a selection of other books follow. A complete list of Oxford Paperbacks, including The World's Classics, Twentieth-Century Classics, OPUS, Past Masters, Oxford Authors, Oxford Shakespeare, and Oxford Paperback Reference, is available in the UK from the General Publicity Department, Oxford University Press (JH), Walton Street, Oxford, OX2 6DP.

In the USA, complete lists are available from the Paperbacks Marketing Manager, Oxford University Press, 200 Madison Avenue, New York, NY 10016.

Oxford Paperbacks are available from all good bookshops. In case of difficulty, please order direct from Oxford University Press Bookshop, 116 High Street, Oxford, Freepost, OX1 4BR, enclosing full payment. Please add 10% of published price for postage and packing.

SHOULD THE BABY LIVE?

The Problem of Handicapped Infants

Helga Kuhse and Peter Singer

This book concerns itself with both the practical and philosophical issues of allowing certain handicapped babies to die. The authors' conclusions are not absolutist in any direction: rather they believe that careful rational examination of the admittedly complex issues—which they present clearly and by way of actual examples—can lead to responsible decisions of different kinds in different cases.

Studies in Bioethics

MORAL DILEMMAS IN MODERN MEDICINE

Edited by Michael Lockwood

Test-tube babies, surrogate mothers, the prescribing of contraceptive pills to girls under sixteen—these are some of the moral dilemmas in medicine that have recently hit the headlines, and will continue to do so for some time. Mary Warnock, Bernard Williams, and R. M. Hare are among those who tackle the ethical problems in modern medical practice from the standpoints of philsophy, medicine, and the law. Other contributors are: Michael Lockwood, Ian Kennedy, J. A. Muir Gray, Raanan Gillon, and Roger Higgs.

THE YOUNGEST SCIENCE
Notes of a Medicine-Watcher

Lewis Thomas

In what does the practice of medicine consist? What is it that people have always expected from doctors? What can they expect now, when medicine is a real science and the old art barely discernible? These are the central questions which Dr Thomas considers as he tells the story of his professional life.

'Apart from its intrinsic literary merit, Thomas's new book has a particular historical significance because the span of his life . . . matches a period when medical practice underwent the most rapid and radical change in its history.' *New Scientist*

'*The Youngest Science* should appeal to a wide readership, not only as the best account available of the present state of medicine, but also for the picture it gives of a remarkable and interesting member of the profession.' *Times Literary Supplement*